BIOLOGICAL
RHYTHMS

The MIT Press Essential Knowledge Series

A complete list of books in this series can be found online at
https://mitpress.mit.edu/books/series/mit-press-essential-knowledge-series.

BIOLOGICAL RHYTHMS

DANIEL B. FORGER

The MIT Press | Cambridge, Massachusetts | London, England

The MIT Press
Massachusetts Institute of Technology
77 Massachusetts Avenue, Cambridge, MA 02139
mitpress.mit.edu

The MIT Press would like to thank the anonymous peer reviewers who provided comments on drafts of this book. The generous work of academic experts is essential for establishing the authority and quality of our publications. We acknowledge with gratitude the contributions of these otherwise uncredited readers.

This book was set in Chaparral Pro by New Best-set Typesetters Ltd. Printed and bound in the United States of America.

Library of Congress Cataloging-in-Publication Data is available.

ISBN: 978-0-262-55314-8

10 9 8 7 6 5 4 3 2 1

EU Authorised Representative: Easy Access System Europe, Mustamäe tee 50, 10621 Tallinn, Estonia | Email: gpsr.requests@easproject.com

To Deborah, Daniel, and Olivia

CONTENTS

SERIES FOREWORD

The MIT Press Essential Knowledge series offers accessible, concise, beautifully produced pocket-size books on topics of current interest. Written by leading thinkers, the books in this series deliver expert overviews of subjects that range from the cultural and the historical to the scientific and the technical.

In today's era of instant information gratification, we have ready access to opinions, rationalizations, and superficial descriptions. Much harder to come by is the foundational knowledge that informs a principled understanding of the world. Essential Knowledge books fill that need. Synthesizing specialized subject matter for nonspecialists and engaging critical topics through fundamentals, each of these compact volumes offers readers a point of access to complex ideas.

THE SECRET LANGUAGE OF YOUR VITAL SIGNS

Rhythms Abound

All life on Earth adapts to the planet's twenty-four-hour rotation. Ancient civilizations understood the importance of rhythms for planting crops and harvesting. Seven-day weeks existed in Hindu and Chinese civilizations, and as far back as at least the Babylonian periods. In almost all aspects of life, timing is essential. Timing is so fundamental that cells have built-in clocks to time events. We pay thousands of dollars for a Rolex when we have billions of internal clocks, all for free. Our body predicts when food will be available, how much sunlight there will be, and how much activity will be needed. Timekeeping is not abstract; it is practical in a world where the early bird gets the worm.

Walk in the woods and you will see rhythms everywhere. Birds chirp before sunrise. Flowers open at set times of the day. On a walk, signals from the sun or the lack thereof are omnipresent. Fishing to survive, you should try your luck early or late in the day, as the fish will follow the *crepuscular* (activity at dawn and dusk) patterns of the insects they eat. Insects avoid the midday sun. Fish track the insects; we track the fish. But this is just one of the many rhythms present. The seasons change landscapes entirely. Tidal rhythms determine where smaller fish migrate, again determining when bigger fish are active, which dictates when we will be active.

Modern life has changed how we interact with these rhythms. With the invention of contemporary lighting, ever-present food sources regardless of the time of day, alarm clocks, screens presenting us with light at all hours, and work at every possible time of day (shift work), it is almost as if we modern humans live on another planet than our ancestors. This change has happened over the past one hundred years or so, since the industrial and digital revolutions—a blink in the eye of evolutionary time. The maladaptation of our rhythms to modern life may explain many modern health ailments.

Medicine cannot fix our rhythms in any universal way. No pill will ever be developed that can be taken whenever we want to fix rhythms. Caffeine does not help when we want to go to bed, but it can help when we want to wake up. *Melatonin* does not help when we want to wake up, but it can help when we want to go to sleep. As we will see, the difficulty is mathematical. Any potential fix of a rhythm at one time of day would worsen the rhythm if the fix were applied at another time. Light in the morning may bring your body closer to European time, depending on where you live. That same light signal, applied in the evening, might bring your body closer to Japanese time.

No matter how much food is present, how many miracle cures are discovered, or even how many new media sources are offered that can better connect us, we cannot escape the rhythms of our body. Understanding these rhythms can help us live healthier and more productive lives. We can time our hours to optimize peak performance, have deeper and more restorative sleep, lose weight, avoid diabetes, run faster, learn quicker, become more fertile . . . Moreover, each person's rhythms are unique to them.

Daily rhythms, which are often called circadian (circa meaning about and dian meaning a day), play a special role in timing biological rhythms. For example, in your body they time the release of melatonin, a hormone that helps us sleep, just before bedtime. Hormones like cortisol are released around when we should wake up. These clocks also cause us to have peak attention later in the morning or peak strength later in the afternoon.

It has never been easier to measure human biology rhythms, such as temperature and heart rate, continuously and in the real world. Sensors can be worn (wearables) and provide real-time measurements of what is happening within the body. Wearables can measure heart rate, temperature, and other rhythms in the real world, over long periods, and in large populations.

Sometimes, they provide physiological data when surveys have been used clinically. The amount of data now is just unthinkable. The secrets to your rhythms live within these datasets.

My research group at the University of Michigan has been obsessed with rhythms for the past ten years. Although I am not a medical doctor (instead I'm a doctor of mathematics), for years my office has been strewn with papers showing these rhythms in sleep, temperature, heart rate, melatonin, mood, the electrical activity of the brain, and glucose. Our database contains over a million days of data from participants in studies run by our collaborators. We have been to many conferences where these rhythms are studied and debated. What these rhythms mean is emerging from these large datasets, academic papers, and debates. My hope is now to bring this knowledge to you.

After parsing through much data and using mathematics like a microscope to see deeper into the rhythms, my collaborators and I have found many rhythmic patterns. These rhythms are related and can be understood at a high level with little background knowledge. Moreover, there are many similarities between the rhythms. All are vital signs indicating the state of physiological systems in the body.

Wearables can measure heart rate, temperature, and other rhythms in the real world, over long periods, and in large populations.

For a more detailed description of our mathematical tools, please see my other book, *Biological Clocks, Rhythms, and Oscillations: The Theory of Biological Timekeeping*, also published by the MIT Press. Other texts are available, as listed at the end, giving the history of discovering biological clocks, particularly *circadian rhythms*, which is so extraordinary that the 2017 Nobel Prize was awarded for this work. We focus on understanding rhythms, which often have signals from internal circadian clocks and other superimposed signals. Here, my goals are simple.

First, I want to show you how to read a time course measured from your body, be it melatonin, heart rate, or any other signal, to learn about the inner workings of your body, especially your biological rhythms. Such a reading could improve health and productivity, but because knowledge is essential here, I do not limit myself to markers used in hospitals or conventions of laboratory researchers. I present what we consistently see in these rhythms in the real world. I provide frameworks to understand each of these rhythms.

My second goal is to show you that the real world and modern life are crucial in these rhythms. What happens in an isolated laboratory or an intensive care unit (ICU) differs greatly from what happens in the real world. The real world can be seen in our physiology. Of course, the basis of our understanding of the physiology of these rhythms is carefully controlled laboratory studies, for which there is no substitute. Likewise, clinical validation is needed before the medical community can adopt any medical treatment. Looking at so many rhythms, however, it is clear that much is left to be understood about what happens in modern life and how this impacts our physiology, which was built for entirely different (i.e., natural) environments. These same systems must work in athletes and sedentary individuals as well as in a college dorm, cabin in the woods, and assisted living facility. They must work for the shift worker and schoolteacher. They must work in rural Haiti and the fanciest resort in San Juan, just an island away.

Lastly, any discussion of rhythms will be based on ideas from statistical, mathematical, or computational algorithms since some method is needed to extract features (e.g., amplitude, period, or phase) from these rhythms (see figure 1). While our focus is not on mathematics, in the age of machine learning (ML), educated readers need to understand some implicit assumptions in these techniques. This way, readers will not be dismissive or overly trusting of these algorithms.

Other texts focus on losing weight, or what happens in lower organisms like flies, fungi, or bacteria. Many of the ideas presented here can be easily translated to behaviors seen in other species. For consistency, though, I needed to pick a model organism. For biased reasons, my choice is the human.

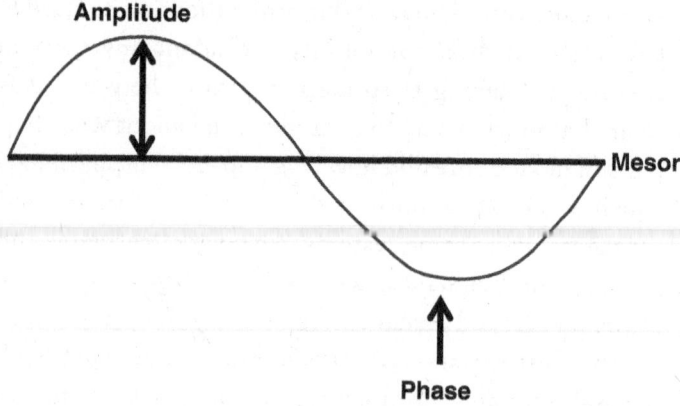

Figure 1 The basic features of a rhythm are its amplitude, phase (which here is defined as the lowest point of the rhythm), and average value (mesor).

There are many situations where understanding your rhythms can be useful. Your watch might predict fertility based on an unusually high heart rate based on the rhythms it measures. If you try time-restricted eating or worry about how shift work can lead to a higher risk of diabetes, your choice of the timing of meals will impact your health. Tracking patterns in your mood may be helpful, for example, if you have bipolar disorder. Irregular sleep rhythms can affect your mood and relationships. Tracking brain rhythms during sleep might give early indications of neurodegenerative disease. Rhythms in temperature or heart rate might indicate that you might have COVID or another respiratory illness based on patterns in heart rate rhythms. . . . I hope this text will help you better understand what is rhythmically happening inside your body in these scenarios. I hope you will also better understand the physiological datasets that devices are collecting from you, which will only grow in size in the coming years.

Privacy with Wearable and Other Time Course Data

Millions of individuals use wearables at the time of this writing, generating the largest physiological dataset in history. The reader might find themselves the owner of years' worth of data that their phone computer or perhaps even a doctor can access. Even more of this data is hidden behind the corporate firewall and will remain inaccessible to the public in the foreseeable future. Learn about these datasets and we learn about ourselves in the process.

We should have access to our data. Others having access to it could raise concerns. Could they read our signals and learn something we want to keep private? For these reasons, all the data we study here, or in my group, are de-identified, meaning we do not know who it is from and if it was shared with their permission. In my research group's apps, we interact with users anonymously. Yet some companies do not adhere to such strict policies. Further discussion of privacy can be found in the last chapter.

How Much Data?

Many projects I now work on easily use 1 gigabyte or more of data, and were I just to read off these numbers, it would take a lifetime. Here is a quick example of the scale of the data. One year of minute-by-minute heart rate and step data is $2 \times 365 \times 24 \times 60$ measurements, or over a million integers. The computer that I write this on can store about 100,000 years of data. Several companies now collect similar data from millions of individuals.

Methodologies are more important than the quantity of data when solving some problems. Unfortunately, in most cases the amount of new knowledge does not linearly scale with the total amount of data. The returns for that extra data can diminish as one gets more data. The data all looks the same past that point, at least at the population level. The point at which this happens is dependent on the dataset. Thus the art of big data science is to know when this happens so one does not leave essential knowledge undiscovered and waste time chasing goals that more data will never achieve.

Wearable data from the real world is not collected as cleanly as in carefully controlled laboratory conditions. The device manufacturers also do not collect data with the same rigor as medical devices, but more important, real life offers many more possible scenarios than those in the lab. So sometimes, unexpected things creep into the datasets. Here are some of the other strange things I have found in wearable datasets:

1. Increased heart rate (> 100 beats per minute [BPM]) for hours without any increases in activity in otherwise healthy adults

2. Steps recorded when an individual was asleep (and likely not sleepwalking)

3. Data collected at erratic intervals, including large (hours) gaps in the data

4. The exact heart rate measurement repeated over many minutes—much more than could occur by chance

5. Heart rate is measured only to about ten BPM (and sometimes less)

The first two instances likely reflect real-world activities (such as bicycle riding). Individuals choose when to wear a device. Companies can decide when to collect data (example 3). And sometimes, one has to be careful about inaccurate measurements (examples 4 and 5).

There is a balance between the amount of data needed to be collected and the quality of the data. Since it often is not the total amount of data that determines the amount of knowledge to be gained, and most devices use similar hardware and technology, what remains are the algorithms.

ML offers many tools to analyze datasets, as we will see from wearables. The success of ML is hard to deny, and this will be an important tool in our tool kit. Nevertheless, it alone cannot be the solution to understanding rhythmic data in the body. Rhythms cannot be captured in a generalized "black box" way, at least with great predictive value.[1] ML alone and unexamined can lead to bias too.[2] ML models are generally uninterpretable in terms of their physiology since, as we will see in chapter 8, their "neurons" are just mathematical functions that behave differently than actual neurons.

Moreover, ML approaches can likely be improved on significantly with knowledge of the physiology and mathematics of rhythms. Thousands of brilliant and careful researchers with gold standard tools have carefully considered physiological processes like heart rate. To ignore their work is like studying motion without the work of Isaac Newton or those who followed him. So ML is helpful, but our past knowledge of physiology is also essential.

Measuring Physiological Signals

Devices will come and go. At the beginning of the wearable revolution, it once seemed as if Jawbone would be the major player for the foreseeable future, but it is now out of business. Fitbit, the next dominant player, was bought by

Google and has now lost market share to Apple and Samsung. This will surely keep changing, and so will technology. The reader will likely see some of these words in the future as we now see Marcel Proust describe an airplane or watch a 1980s' sitcom showing reams of paper emerging from a dot matrix printer.

Still, the physiological systems that the devices measure change over a much longer evolutionary time. The fundamental methodologies, the mathematics, improve on the timescale of decades or perhaps even longer. For this reason, we spend relatively little time describing devices, and instead focus on physiology, data, and analytic methods. These methods will be relevant for many years to come.

Nonetheless, it is helpful to briefly overview the devices currently used to measure physiological rhythms in the real world. Users typically assume that devices accurately measure what they say. Yet the details of what constitutes a measured step or sleep bout and so on, are private, patented, or a company secret, so we do not fully know what is being done. The algorithms that determine a step, for example, can also change quickly as a company updates its app. An algorithm can only be fully understood and validated if it is disclosed. For this reason, readers should prefer algorithms that are "open access."

We are concerned with rhythmic patterns of what is measured. How does heart rate vary across the day? How does sleep or mood vary across seasons? So much of our discussion will take the methods to measure these quantities as correct. This is not necessarily true, however, and sometimes the details of what goes into the measurements need to be considered. Below is a quick primer to get us started.

Chapter 2 revolves around sleep, which is now ubiquitously scored with wearables. Movement during sleep and wake is measured by *triaxial accelerometry* or sometimes angular velocity. The details of the device are not particularly important for our discussion. Other signals, including heart rate, can be used in sleep scoring.

Chapter 3 focuses on hormone rhythms, with melatonin rhythms as the key example. The phase of the melatonin rhythm is typically measured by *dim light melatonin onset* (DLMO). Here, multiple melatonin measurements are taken when an individual is in near darkness, since melatonin levels are sensitive to external light levels, and the time when melatonin begins to rise is recorded. Wearable devices that measure melatonin concentration under the skin are currently being developed.

Chapter 4 explores rhythms in temperature. Sometimes, *skin temperature* is used as a proxy for core body temperature, such as heart temperature. As we will see, though, these two processes could have opposite dynamics. For instance, as the body cools, core body temperature decreases and skin temperature increases as blood vessels bring blood closer to the skin's surface. Both core body and skin temperatures are useful.

Chapter 5 looks at heart rate, which is typically measured via *photoplethysmography*, whereby light is shone on the skin to determine how much has been absorbed. This determines the blood volume through arteries. Other information from this signal could inform blood oxygen levels, blood pressure, and atherosclerosis.[3]

Chapter 6 examines mood. Mood is typically measured through questionnaires such as the *Patient Health Questionnaire* (PHQ-9), which focuses on depression. Mania can be measured via the *Altman Self-Rating Mania Scale*. Many simpler measures exist, however, such as rating mood on a scale from one to ten, or choosing which facial expressions best match an individual's state.

Chapter 7 concentrates on *electroencephalogram* (EEG). These are rhythms in the brain's electrical activity recorded after they have passed through the scalp. As few as three or as many as sixty-four or more electrodes on the head can simultaneously measure this signal.

Chapter 8 focuses on glucose rhythms. Many patients with diabetes use *continuous glucose monitors* (CGMs). These devices continuously measure glucose. Although users rarely consider glucose rhythms, they could be crucial.

Measuring Natural Environments

Since rhythms are often used by the body to anticipate aspects of the environment, separate from measuring rhythms, one may need to measure light levels, external temperature, pressure, and position (GPS) in the environment. These environmental variables affect physiological systems and can help us predict their state. They will be important considerations in our discussions, but not our focus as they are not physiological measurements.

Similar to physiological measurements, one might need to look carefully at the details of environmental measurements to know what is truly being measured. For instance, light affects much of our physiology. It does this through our eyes, yet most light measured by wearables is light measured at the wrist, which can be covered by sleeves and pointed in directions separate from the eye. So data from a study where light was measured at the wrist might need to be discounted compared to light measured near the eye.

Here is another example: to save energy and bring our campus closer to being carbon neutral, the lights in my office were changed. While they emit similar amounts of light to the previous ones, the color spectrum of the new lights is much closer to what one would find at dawn or dusk. The key decision-makers probably missed the spectrum of the light, but these new lights will cause decreased alertness. The brightness of the light may not be the only thing we need to measure to understand environments.

Likewise, social factors affect many aspects of our environment. Our social environments impact how we sleep, our heart rate, and our overall health. Quantifying

What we measure may depend on where and when we measure it.

environments, such as whether they increase stress or fear, may also be important. What we measure may depend on where and when we measure it.

Applications of Our Work

So what are the benefits of tracking rhythms? There are many possible applications, some of which will be described in depth later. Here are some critical applications of rhythms research to give you a sample of what is to come. These are taken from the work of my collaborators and me to give you more of a personal flavor of the field.

Chronomedicine

As we will see, much of our physiology varies based on the time of day. The vast majority of the targets of approved drugs show rhythmicity. When clinical trials looked for timing effects, 70 percent of the studies found them.[4] Our metabolism shows prominent twenty-four-hour rhythms. This means that drugs taken at certain times of the day will be more effective than others. These drugs might more slowly degrade or target a system when it is most active. Drugs that are quickly metabolized, however, could have significant daytime or nighttime effects.

Even the simplest cases of how rhythms affect drug action still need to be understood widely. For example, I have a bottle of melatonin pills that I sometimes use to help me when I travel overseas. But the bottle does not mention anything about taking melatonin at night, even though we will see that it signals the biological night. Melatonin can also shift timekeeping in many parts of our body and help us fall asleep, but these claims have yet to be fully validated by our medical system. Plus with so much variation between individuals in how much sleep is needed and the timing of it, it is unlikely that any one set prescription will work for all.

Still, in addition to the direct effect of drugs, rhythms affect the whole medical system. Fatigue, which is closely linked to medical errors, depends on sleep deprivation and circadian rhythms in alertness. The timing of surgery can also influence its outcomes. This could be due to medical error, but more likely our body's ability to recover and tolerate injury depends on the timing of the surgery or injury. When one takes vaccines impacts their ability to prevent future disease. So rhythms are important for disease prevention, medical errors, side effects, and a host of other factors that affect our ability to heal.

Heart Attacks
On waking from sleep, the body releases many hormones to counteract the effects of sleep and bring us to full attention. This is a time when heart attacks are more prevalent too. Molecular clocks within heart cells can directly control how excitable the heart is. Could this explain the

timing of heart attacks? One critical link is the potassium channel interacting protein 2 (KChIP2).[5] KChIP2 is a protein whose expression levels vary across the day. When the heart beats, kChIP2 extends the amount of time the heart cell's voltage is elevated, leading to stronger heartbeats. It also can cause individuals to be susceptible to cardiac arrhythmia, where electrical signals do not correctly propagate through the heart.

This example illustrates how rhythms on different timescales are intertwined in the human body. The heart beats on the timescale of a second. Activity, on the timescale of minutes to hours, can speed or slow this down to meet metabolic demands throughout the body. Proper metabolism is essential to fueling the heart, and rhythms in metabolism might damage the heart. Likewise, daily metabolic rhythms cause the heart to beat faster or slower, thereby affecting electrical signals' ability to spread throughout the heart.

The Apple Watch now warns us if our heart rate is irregular. It can even record an electrocardiogram (ECG). By understanding these rhythms better, we can instruct individuals *when* to put the watch into ECG mode and better detect irregular heart patterns. A better understanding of these rhythms can help us determine what is going wrong.

Help for Shift Work

Shift workers are on the front lines of the fight with their biological clocks. These heroic workers who keep our power plants operational, military ready, and hospitals open expose themselves to significant health risks. Of these workers, the most affected are the permanent night shift workers. Permanent night shift workers are in a state of constant flux. During their shifts, they sleep during the day, only to try to align to regular days during weekends and between shifts to attend children's birthday parties, renew their driver's licenses, and connect with most of society.

As we will see, such workers' bodies are in a state of disorientation. Their hearts could be in one time zone while their livers are in another. The body might secrete melatonin, the hormone that tells the body it is night, just when they need to perform at peak performance. The physiological systems in their body hear many conflicting signals and might mistime events. The long-term consequences of shift work are so significant that it is now widely believed to increase the risk of cancer, diabetes, and mental health disorders.

Apps can determine what time the body thinks it is from wearable data, and then give shift workers suggestions about when to time meals, light, and other aspects of daily life to achieve goals such as being more alert during shifts, sleeping better, or better participating in activities during their days off. We will explore many of the rhythms that can be misaligned in shift workers and what this misalignment will mean for overall health.

Performance

I realized how important timekeeping is when I saw a plot of the frequency of deadly accidents at different times of the night.[6] Even when scaled for time awake, fatal accidents are much more common in the middle of the night when our brains are prepared for sleep. This extends to other kinds of errors. So the next time you are about to undergo surgery, ask your doctor how long they have been awake. Feel free to ask a taxi driver that question before getting in the car.

One fascinating aspect of performance is its interindividual variability. Not only are some individuals better at some tasks than others but some are also particularly good at performance at the wrong circadian times or in the presence of sleep deprivation, perhaps because of genetics.

I am now leading a multi-institutional project funded by the Army Research Laboratory to get to the biological basis of cognitive fatigue. This fatigue affects our mental tasks, such as driving a vehicle. Cognitive fatigue has a strong circadian component and depends on how long an individual has been awake. But many rhythms of cognitive fatigue still need to be fully understood. For example, there is a rhythm of performing a task and then needing a break. Performance shows great interindividual variation, with some individuals being particularly resistant to performance deficits due to sleep deprivation for reasons we do not fully understand.

Cancer

Throughout this text, you will see many references to the work of oncologists. Rhythms are essential for understanding cancer. At the highest level, cancer is a disorder of uncontrolled cell division. Cells divide in a well-understood rhythm. Before they divide, checkpoints determine whether to proceed; cell growth and division should only occur for healthy cells. In addition, cells only divide at certain times of the day, as determined by their internal clocks. Amazingly, at least for some cancerous cells, cell division occurs three times a day, roughly eight hours apart.[7] So once again, we find another disease at the intersection of multiple timekeeping systems.

To further highlight the role rhythms play in cancer, consider the timing of chemotherapy. In remarkable studies by Francis Levi, chemotherapy at different times of day can drastically affect survival rates. Yet many of the details remain to be determined. For instance, the times of day best for chemotherapy in men might not be the same as in women.[8] Additionally, the timing of chemotherapy might have more to do with how much can be tolerated rather than its ability to kill cancer cells at particular times of the day.

One of the biggest concerns for cancer survivors is fatigue. I have been part of a large effort funded by the National Institutes of Health to develop an app to track the rhythms of fatigue in cancer survivors. The app also suggests ways to increase the amplitude of these rhythms, allowing better sleep at night and less fatigue during the day.

Decoding Rhythms

These applications are just a few potential benefits of understanding the body's rhythms. Most applications still need to be discovered. But before making these discoveries, we must understand these rhythms in the real world. With wearables, you can now have access to your data from rhythms in many systems of your body. Rhythms, oscillations, are mathematical objects, but our goal here is not to dive deep into mathematics. It is to provide you with a better understanding of these rhythms based on the past mathematical and biological basic science that has been discovered. Our aim for the following chapters is to help you understand your rhythms so that you can use them to lead healthier and more productive lives. The chapters can be read independently, so you can jump to the discussion of the rhythms you are most interested in.

SLEEP

What if I were to tell you that you would collapse, be transported to faraway worlds in the future and past, and remember almost none of it? Each night, we find ourselves in this unique world during sleep. One-third of our lives are a mystery.

Even though many of us try, sometimes remarkably, there is no way to avoid sleep. My seven-year-old daughter insists she is not tired, yet I find her asleep two minutes later with a book covering her face. While visiting a lab in Japan, I saw a researcher seemingly focused on their work with two hands holding instruments. They were not moving. At a glance, they were likely working. Yet they were asleep and can stay in that position for hours. Or the person on a California freeway "driving" their Tesla home, yet appearing to be asleep. I, too, am guilty of ignoring my sleep debt. On a one-hour commute on the subway each way to high school,

I somehow found just the correct position to sleep through the sensory overload of NYC transit, yet somehow would almost always wake just before my stop.

Trying to rob the sleep bank can have devastating consequences. What would have happened if the staff at the Chernobyl power plant had decided to conduct tests in the afternoon rather than in the middle of the night? Could one of the many mistakes required for the nuclear meltdown have been avoided? It seems likely to this author. Similar accidents have occurred in all countries and cultures. More recently, the crash of the USS *Fitzgerald* was linked to operator fatigue. That crash caused loss of life, damage in the hundreds of millions of dollars, and congressional hearings.

Lack of sleep can have severe health consequences, just as too much sleep might indicate poor health. Not getting enough sleep can lead to cognitive decline. As we will see in chapter 6, poor sleep is a feature of and perhaps exacerbates psychiatric disease. Less sleep means a higher risk of getting an infection. This partially explains why I (and others) often get respiratory infections on overnight flights despite the recent improvement in cabin air filtering. Taken together, this has caused many to think that a pandemic of sleep loss has aggravated modern pandemics.

Millions of individuals track their sleep with wearables. A boon to major tech companies including Apple, Google (Fitbit), and Samsung, it is unclear what individuals will do with this information. Sure, I slept only four hours last night, but what does that mean? Is it healthy? When would be best to sleep next? When should I nap? What does the "restorative" sleep calculation on my phone tell me? This is harder to understand because rhythm lies at the center of it. But with some explanation, you can understand sleep rhythms better.

Rhythms can explain some of the mysteries of sleep. There is no way I know of to avoid sleep, but there are ways to sleep better. Sleep is more effective at certain times than others. Moreover, we are just more alert at certain times than others. So understanding the rhythms of sleep can greatly improve quality of life. And they are fascinating. They beat, pulse, and dance. We just need to get them to beat to the right tune.

Homeostatic Rhythms

Stay awake and you get tired. Sleep, and this tiredness dissipates. This is the most fundamental rhythm of sleep. This rhythm is more relaxed than in a wall clock. Some days, we sleep more. Some days, we sleep less. Our bodies keep account of sleep. If we sleep less one night, the next night we tend to have more sleep.

An extreme example of this is the sleep patterns of college students. During the week, sleep can be minimal.

Understanding the rhythms of sleep can greatly improve quality of life. And they are fascinating. They beat, pulse, and dance. We just need to get them to beat to the right tune.

Some nights might have little sleep. Many students make up for this lack (sleep debt) by sleeping in during the weekend. But this debt is not strict. Suppose we sleep only five hours every day for a week. Given the standard eight hours of sleep suggested by some, we will have accumulated three hours a day or twenty-one hours of total sleep debt over the week. Now try to recover from this sleep debt. It is doubtful that you will sleep for twenty-one hours straight. Ten hours would be a feat for many. So where does this sleep debt go?

The key to this paradox is a mathematical function called an exponential, perhaps the greatest mathematical gift humanity has ever received. It is responsible for the compound interest that allows many of us to retire, but it also explains why some pandemics rage out of control. With the exponential function, be it the number of infected individuals in a pandemic or dollars in a bank account, the rate of change depends on the current amount. As we get more money, we make more money. That makes it powerful and causes values to proliferate. The effects of sleep debt are like an exponential in reverse. Its accumulation slows with growing sleep debt, and its payback decreases with more sleep. This allows us to stay awake for a day or two and still be functional.

When we first fall asleep, we pay back our sleep debt quickly. With every hour of sleep, the amount of restoration one receives decreases. This lets us sleep deeply at first

The effects of sleep debt are like an exponential in reverse. Its accumulation slows with growing sleep debt, and its payback decreases with more sleep.

and recover as much as possible. Also, as we stay awake longer, we accumulate less sleep debt than we did earlier, perhaps because some brain functions might start to act as if they are asleep (causing performance deficits). In this way, the need for sleep and the effects of sleep do not accumulate linearly. As we will see later, our ability to judge fatigue is flawed, and few of us fully understand this sleep exponential.

What is happening is that as we stay awake longer, our brains are slowly shutting down. The brain is learning at slower rates and paying attention to fewer signals. Sleep deprivation can cause serious consequences, such as drivers missing a car swerving into their lane, or even less serious but important consequences too, such as diners making errors in calculating tips at a restaurant. Taken together, while staying awake longer accumulates more sleep debt, it does so at a lower rate. Sleep for longer yields less restorative ability, allowing us to draw from a sleep debt overdraft and never have to pay it back. I wish that were the case in real life.

So some "clever" individuals sought to take advantage of these properties of sleep. If the beginning parts of sleep are the most restorative because of the sleep exponential, could we just sleep for a few hours, get the most restorative parts, and add hours to the waking day? Taking naps is OK, but these naps should not substitute for decreased overall sleep. The famous mathematician Norbert

Wiener claimed to only sleep for a couple hours at a time; for most, however, this has severe consequences and may even trigger psychosis. My strong advice is to avoid this polyphasic sleep like the plague. This is because we do not truly understand what sleep does, and it likely performs several different functions, such as clearing toxins from the brain, muscle repair, and organizing memories, and even dreaming plays functions we do not fully understand. While pulling an all-nighter or occasionally having an irregular sleep pattern can be dealt with, it should not always be adopted.

Not all tasks are the same, and not all generate sleep debt as quickly as we would or would not like. If counting sheep quickly accumulated sleep debt and intense concentration could occur without much cost, we would be set. Unfortunately, many things cause increases in sleep debt, and they can act in unintuitive ways. For example, staying awake for a long period causes sleep debt, and so does fighting an infection, running a marathon, or intense concentration.

Plus we may feel less awake early in the morning and more tired throughout the day. Many experience "second winds" or other unexpected bursts of energy, which as we will also see, are due to circadian timekeeping. In addition, these clocks can explain the differences between fatigue in people who identify as night owls and morning larks.

Circadian Rhythms in Sleep

We typically get more tired the longer we stay awake. So should we *feel* more tired as the day goes along? Should we not feel much less energy in the afternoon than when we wake up? In general, this is not the case. As the day progresses, we continue to have energy. Early evening is a tough time to try to go to sleep. Richard Kronauer called this the "forbidden zone," during which sleep was unlikely. What is going on here? Circadian rhythms to the rescue![1]

Just as we are about to feel the afternoon and early evening fatigue, our circadian clock steps in. It can dramatically increase or decrease fatigue depending on the time of day. Just as we are about to be dragged down by fatigue in the evening, the circadian clock gives us a substantial boost. Our cognitive ability is better than it should be. Our ability to fall asleep decreases. This is less of a boost than a plateau, maintaining our ability to keep alert. Yet the circadian rhythm is a rhythm, and it starts to drop off at a certain point (see figure 2). This drop-off is compounded by the increasing fatigue that is building up. Moreover, the clock causes melatonin release. The rising melatonin, increasing fatigue, and decreasing circadian drive direct us to be ready for bed. Before we know it, our spouses hear our snores.

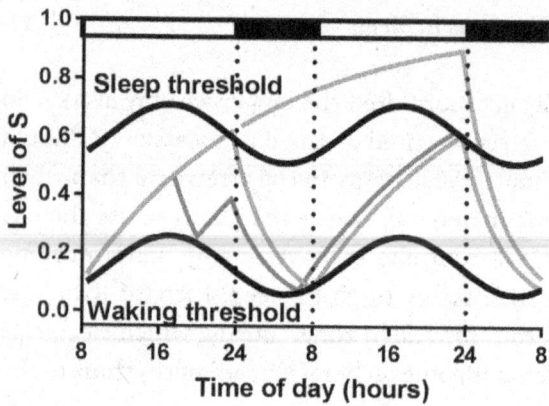

May 18, 2024

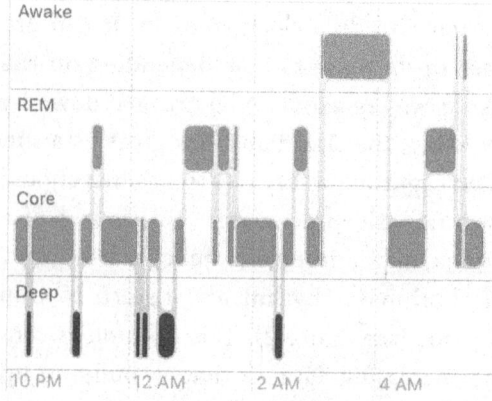

Figure 2 Top: An illustration of the two-process model of sleep regulation. Individuals transition between two oscillating thresholds (solid curves) during sleep or wake. Sample trajectories (gray lines) showing normal conditions, sleep deprivation (where the gray curve overshoots the sleep threshold), and a nap (where the gray curve undershoots the sleep threshold). *Source*: Tom Deboer, "Sleep Homeostasis and the Circadian Clock: Do the Circadian Pacemaker and the Sleep Homeostat Influence Each Other's Functioning?," *Neurobiology of Sleep and Circadian Rhythms* 5 (2018): 68–77. Bottom: An example of the author's sleep as scored by his Apple Watch.

Pull an all-nighter and a strange thing happens. Just as our fatigue should be compounding, we get our second wind. Somehow, we can plow through morning meetings and even make it to the afternoon without collapsing despite a day of sleep deprivation. Again, our sleep debt continues to build up at a lesser rate. Plus our circadian rhythm kicks in. Again, circadian rhythms save the day!

You might experience this when crossing time zones. We stay awake on the day of the new time zone. But if the circadian rhythm in *sleep drive* is mistimed, just as the clock should push us to be more awake and survive the day, it might give the opposite drive. Furthermore, even though we have accumulated sleep debt, we cannot fall asleep because circadian alarm bells are ringing. Even if we do sleep, we may wake up shortly afterward. The quality of sleep can be poor as well, and even though we need to pay back the sleep debt, our circadian system puts us on a payment plan, limiting how much we can be restored. Sleep is elusive, and we just are not ourselves.

The lesson from all of this is that understanding fatigue requires understanding sleep and circadian rhythms. Sleep needs to occur at the right phases of the circadian cycle for the most efficient sleep. Naps in the late morning or early evening just are not as productive as at other times. Sometimes the circadian clock is misaligned with the external world (e.g., jet lag) and it is difficult to know when the most efficient sleep would occur. This is described later.

Understanding fatigue requires understanding sleep and circadian rhythms. Sleep needs to occur at the right phases of the circadian cycle for the most efficient sleep.

REM-NREM

Now we will explore more of the mysteries of sleep. Not all sleep, as dramatist William Shakespeare said, "perchance to dream," is the same. Within sleep, there is another rhythm with a period of about ninety minutes. During this rhythm, we alternate between a *sleep stage* where we can dream (rapid eye movement [REM], also known as paradoxical sleep) and deeper sleep, simply known as non-REM (NREM) sleep.

When in REM sleep, the brain looks like when it is awake. There is desynchronized activity in the brain. Desynchronization is good since it allows the brain to process many different signals. Organization, in general, is not a good thing for the waking brain. Epileptic seizures are an example of pathological organization within the brain. While some movement (e.g., REM) occurs, more significant movements, such as walking, talking, and so on, are suppressed. In some cases of REM disorder, these can manifest, but mostly we are glad they are not manifest. We will discuss these rhythms in much further depth in chapter 7.

NREM sleep is a much "deeper" sleep and shows up easily when measuring the brain. During NREM sleep, EEGs, which measure brain activity by recording small electrical signals that can be detected by electrodes placed on the scalp, show slow rhythms as slow as a second in a period—much longer than most other EEG rhythms. These rhythms contain some downtimes when the neurons in the brain are silent and inhibited, almost as if the brain shuts down briefly. This allows the maintenance systems of the brain to work, clearing toxins and removing connections that might be so strong that they cause damage.

Sleep is typically considered more restorative if it contains more NREM sleep. But again, one has to be careful of such claims. We need to fully understand how much NREM sleep is needed. Different individuals can get almost double the amount of NREM sleep as others. Some of this is genetic, as seen in the last chapter.

Scoring Sleep from Wearables

Much of the recent interest in sleep has been sparked by recent work to classify sleep based on wearables. To understand this recent interest, we consider how wearables score sleep. The key experiments involve having someone spend a night in the sleep lab, where sleep is scored by gold standard methods called polysomnography, which measures signals like EEGs, breathing rate, heart rate, and motion from the body during sleep. From these signals, a subject can be classified as awake, in REM sleep, or in NREM sleep every thirty seconds. We repeated these experiments, except we also had the subject wear a wearable.

Techniques from ML can then use the wearable data to predict the sleep state and compare predictions against the classification from gold standard polysomnography data. The most used wearable signal is triaxial acceleration (measured motion), but heart rate plays a key role too. In this way, my collaborators and I developed the first open-access algorithm to score sleep from the Apple Watch. Since the algorithm was open access, a short time later, Apple started to score sleep from the watch. We will never know how much our algorithm influenced theirs. We are just happy that it might be used.

There is a fallacy in terms of sleep scoring. The main goal of sleep scoring is to determine if an individual is asleep or awake during the normal period when they should be asleep. Modern algorithms boast accuracies of 90 percent, which sounds good and sells devices. Here, researchers define accuracy as the number of correct sleep predictions or number of predictions. But consider the simplest mathematical algorithm that predicts that a person is always asleep during this period. Since individuals typically sleep for about 90 percent of this period, this simple algorithm achieves 90 percent accuracy. In this way, a 90 percent accuracy algorithm does not seem innovative.

Beyond accuracy, the developers of these sleep-detecting algorithms rely on the impressiveness of their sensitivity to advertise their products. Sensitivity, in this case, is the ratio of predicted time asleep to actual time asleep. A bigger challenge for such sleep-scoring algorithms is how often they correctly predict that a person is *awake* during the normal sleep period. This is called specificity, and this example measures predicted time awake versus the actual time awake. For now, anything greater than 50 percent accuracy and specificity is good for the sleep-scoring algorithms, although the details of how a specific ML algorithm works are difficult to interpret, even for ML experts. In this simple algorithm, more motion can predict that someone is awake. This is not a perfect algorithm for scoring REM sleep, which needs information from heart rate or other vital signs. New algorithms are being developed, however, and the expectations for accuracy and specificity will increase.

Scoring sleep becomes much more difficult if we try to distinguish between REM and NREM sleep. We can use motion, but heart rate is a better predictor. This is because motion is suppressed during sleep, but variations in heart rate can persist in NREM sleep. The ability to predict different kinds of sleep from a wearable is nevertheless much different than if someone is awake or asleep. For these reasons, many sleep medical doctors I have spoken with tend to ignore the scoring of NREM versus REM, deep versus light sleep, and so on, typically reported by wearable devices.

We previously discussed how circadian rhythms were essential in understanding sleep. Measuring circadian rhythms also improves the scoring of sleep. In particular, some algorithms use the predicted circadian rhythm in sleep drive, which could be determined separately from wearable data to improve prediction. Since knowing the state of the circadian clock (or at least the predicted state) helps classify sleep, it provides strong evidence that the circadian pacemaker's drive on sleep plays a vital role in sleep timing in real life. What is further needed are algorithms and apps to let us know how likely it would be for us to sleep at a given time. This could help us time our nights.

Sleep Mechanisms

The past ten years have yielded tremendous information about sleep science. While we do not fully know why we sleep, we know much more about why we might need to sleep and where its rhythmicity comes from. The prevailing hypothesis about the reason for sleep was proposed by neuroscientist Giulio Tononi and colleagues, and is called the *synaptic homeostasis hypothesis*.[2] They suggest that *synapses*, the connections between neurons, are strengthened during the day. This occurs as more information is provided to the brain and learning proceeds. Yet these excitatory synapses can only strengthen so much. Too much signaling and the neurons would fry. More scientifically, stronger synapses cause more calcium to flow into a cell. While calcium is the most important signaling molecule within the cell, too much calcium causes cells to die. Since neurons in general are not replenished, dying neurons are a grave matter. Each night, we need sleep to restore synaptic strengths to workable levels. All the sensory information we learn is shut off, and the brain can prune back the synaptic connections that are not needed, storing key information in memory while discarding unnecessary information. This creates a rhythm in synaptic growth and decay where the increase and decrease in sleep drive are directly related to the number of excitatory synapses.

György Buzsáki and colleagues, notably fellow neuroscientist Brendon Watson, challenged this hypothesis somewhat.[3] They observed that the signaling between some neurons decreases during sleep, at least in that the neurons do not "fire" as often. But they also saw the opposite happening. Some neurons that were silenced during the day start to increase their firing during sleep. Sleep homogenizes the signaling between neurons and sets them in a more neutral state. Thus a rhythm is created where learning during the day can cause neurons to signal either more or less, and sleep restores neurons to their baseline signaling. Further work is needed to distinguish between the Tononi and Buzsáki–Watson hypotheses, and explore these suppositions in many different brain regions.

Another hypothesis has been proposed with the key signaling molecule, adenosine. As we stay awake, adenosine rises, because of both brain and muscle activity, and it decreases during sleep.[4] It does this in a way that seems proportional to sleep need. Caffeine, which we know has a significant effect on sleep, so much so that it is the most used drug in the world, affects adenosine receptors, tricking the body into thinking there is less adenosine around and hence perhaps making us feel less tired. Some holes in this theory also need to be patched, though. Animals that do not signal with adenosine can still show normal sleep, which seems counterintuitive to the idea that adenosine regulates sleep.[5]

More recently, growing evidence suggests that sleep is essential to clear out harmful chemicals and signals that accumulate in the brain during the day. Like the lymphatic system in the rest of the body, the brain has one called the *glymphatic system* that clears out harmful substances during sleep. This system could also be at the center of why we sleep. How this system works during sleep is still being worked out.

Synapses, adenosine, and even inflammatory molecules (which supports why we get sleepy when sick) are likely players in cognitive fatigue. There are different types of cognitive fatigue corresponding to these separate factors. We might be too tired to do some tasks but able to perform others. Such mechanisms likely provide the biological underpinnings of the rising and falling exponentials that describe the effect of sleep deprivation.

Naps

The restorative power of naps is remarkable. Give up thirty minutes or an hour of your conscious life, and you can feel better. Plus taking a nap for a shorter period, like thirty minutes, can restore your performance without the grogginess (the technical term for this is sleep inertia) that occurs after a long sleep bout. Try it. Put this book down for a thirty-minute nap.

Naps have a clear rhythmicity that we should consider. Napping in the afternoon is often much easier than at 10 a.m. Moreover, napping at 6 p.m. could interfere with the ability to sleep later on at night. The cause of this remains unknown. Here I will argue that this napping ability is hardwired based on a man who is over eighty years old.

One of the most interesting sleep records I have seen is of an older man brought into the clinic. Like many individuals his age, he had a clear pattern of nighttime sleep and a robust afternoon nap. When he was studied in the sleep lab, all contact with the external world was removed so that he had no hint of the time. He could not speak with his family. No watches or clocks were allowed in the room with him. Scientists even set the lights to a constant dim level. He was free to sleep as he wished. Over a week or two, a fascinating pattern emerged. His nighttime sleep bout shrunk, and his nap bout grew. Both went out of phase with a night in the real world (for which he had no clue). When he returned to the real world, the "nap" bout drifted in phase with the world's night and became his nighttime sleep bout.[6] This one individual suggests to me that the timing of naps likely relates to a similar mechanism that regulates the timing of nighttime sleep. In fact, in this man's case, his naps became his nighttime sleep bout and his sleep bout became a long nap.

We will never know what was happening in this man's head. The closest we can get is to study mice with a remarkably similar circadian timekeeping system to humans. The only difference is that mice are nocturnal, whereas humans are diurnal. Still, this switch occurs after the circadian clock, in that the clock sends the same signals in both nocturnal and diurnal mammals. It is interpreted differently in downstream brain pathways.

My collaborator Steve Brown sought to figure out this science of napping. In flies, there are dedicated cells that time many parts of their daily activity cycle. Could this be true for mammals too? In his lab, Ben Collins explored this and found cells that control napping.[7] Remove them, and mice take fewer naps in the middle of their busy times. Increase the signaling from these cells, and animals nap. Thus the time of napping is regulated by specific cells within the suprachiasmatic nuclei (SCN).

This brings up many key questions. How can these cells control naps? Do they act similarly in nocturnal and diurnal animals? Why do they time napping when many other cells might suggest being awake? Unfortunately, we are set back many years in this research as Steve died in a plane crash at fifty-two years of age. For me, this was more than science; it was personal. Steve was an amazing friend whose love of science was infectious and one of the most generous people I have ever met.

Dance of *Entrainment* on Mars? In Basements?

Eventually, humans will travel to Mars, hopefully within my lifetime. Another challenge remains beyond finding or growing food as well as surviving the atmosphere and temperature gradients. The Martian day is more than one hour longer than our own. To study this, circadian researchers such as Kenneth Wright and Charles Czeisler have had humans live on simulated Martian and other non-twenty-four-hour days, at least in terms of their light-dark cycle. Scientists force sleep patterns into this schedule by preventing sleep during the light period, which can last for more than sixteen hours, and then having live in the darkness for more than eight hours. What happens is a rather remarkable dance of rhythms.

Suppose the light-dark cycle period is close to twenty-four hours, for example, within ten minutes. In that case, humans will typically adjust to this new light-dark cycle, perhaps also waking up later or earlier in this new cycle. This means that given enough days to adjust, eventually their sleep will align with night. On returning to Earth, given enough additional days to adjust, they will align back with Earth days. If individuals try to live in a light-dark cycle that is far from twenty-four hours, such as twenty hours of light and ten hours of darkness, the circadian system cannot adjust, even over a month or more. Their melatonin rhythms will have a near twenty-four-hour period, with the same period as that found intrinsically.

In between these two extremes is a beautiful phenomenon called phase trapping. Here the average period will be twenty-four hours, but the circadian system will advance for a few days to catch up to the imposed light-dark cycle, only to slip back and delay for a couple days. When the external day is far, but not too far, from the internal period of the circadian clock, researchers find that the clock can adjust to the external day.

This then begs the question, What if scientists gave no light to an individual? If so, would we have a rhythm of longer than twenty-four hours? If we could turn on lights whenever we wished (e.g., as if we were living in a cave), we would have a period closer to twenty-five hours.[8] This longer period comes about since light during our biological day (when we turn it on) tends to delay the clock and cause us to have a longer day. Interestingly, even outside a cave, some individuals live non-twenty-four-hour days, such as because they are blind and cannot see the twenty-four-hour light-dark cycle or are isolated (e.g., because they live in a basement and rarely go out). These non-twenty-four-hour patterns are intriguing, but the author cautions the reader from trying them at home.

Sleeping across the World

Several years ago, my research group released an app, Entrain, which helped travelers adjust to new time zones using mathematical modeling. The app went viral and was downloaded over two hundred thousand times. When individuals installed the app, they were asked to self-report when they typically went to bed and woke up. We also had their self-reported home city, which the time zone stamp from the phone data could verify. We realized this gave us much information about how people slept differently in different cities.

We found that individuals in Japan had less sleep than in many other countries, as we would have expected from the popular press. But more surprisingly, we found that women slept more than men on average and pretty much at all ages.[9] One caveat about this research is that it is self-reported, so it was possible that women just reported more sleep than men rather than actually sleeping more than men. Subsequent research, however, has validated our finding with sleep scoring via wearable technology.

How Much Do You Need to Sleep?

Spoiler alert: No simple answer can be given to this question. Sleep timing depends on previous sleep history and circadian rhythms. It depends on what you have done, how much you have taxed your brain and body, and even the amount of light you have received. Genetics also plays a role, as do social factors. We can always stay up late to check that last-minute email or talk with people around us. Yet waking occurs from an unconscious state. Separate from alarm clocks, our biological clocks tend to wake us up.

Some individuals get by with five hours or less sleep, whereas others need more than eight hours to function accurately. Why this occurs remains a mystery, but it must have something to do with sleep depth and quality. An equally crucial unanswered question is how much sleep is needed to perform cognitive tasks, like successfully reacting to another car swerving into your lane while driving.

The amount of cognitive fatigue individuals experience varies significantly between individuals. This variation can be more significant than even the mean effects. So even if your neighbor claims to drive accurately with minimal sleep (likely a false claim), you might have a different fate. Additionally, sleep timing varies greatly. It is not just about the amount of sleep; timing is important too.

Tracking sleep patterns and rhythms is fun, but it can be informative as well. We can add to our knowledge by using wearables and apps to analyze sleep rhythms. I hope that this chapter will help you learn what may have caused your sleep pattern. Ideally, this information could lead to you getting a better night's sleep. Speaking of sleep, time for me to sign off.

MELATONIN

Composer Wolfgang Amadeus Mozart's "Queen of the Night" aria from *The Magic Flute* is a classic in classical music literature. Anything but deferential, it shows off independence and the singer's technique from the highest highs to the lowest lows. For circadian biologists, melatonin, the "hormone of the night," plays a similar role in circadian biology. It keeps its highs and lows regardless of many external factors. Yet light is its Achilles' heel. Small amounts of light can quickly shut down or change the timing of its production for many days. In Smith–Magenis syndrome, melatonin rhythms are inverted and seem to time the day rather than the night.[1]

Melatonin rhythms present an ideal example for understanding the basic concepts of circadian biology, which will be a focus of this chapter. Understanding melatonin rhythms will unlock many key concepts that help to explain other rhythms in the body. This is why we start with a discussion of a hormone you may not be particularly familiar with before exploring, in later chapters, more familiar topics like sleep and mood. We first give some background on melatonin and then illustrate some principles of circadian timekeeping with melatonin.

While we will focus on melatonin in this chapter, biological rhythms in many other internal signaling molecules follow similar patterns. Consider circadian rhythms in dopamine or cortisol. Dopamine is released in the ventral tegmental area. The adrenal glands release cortisol. Their biochemistry is different than melatonin. Yet their rhythmic behavior, such as being controlled by intracellular molecular circadian clocks or signaling from other brain regions, which we will discuss later, follows a standard paradigm. Their rhythms also obey basic principles we will study that apply across all circadian rhythms.

How Is Melatonin Made and Measured?

Melatonin is a hormone released from the pineal gland, located near the center of the brain; we are only just beginning to understand its function. It is produced from serotonin and other ubiquitous signaling molecules. We do not know exactly what melatonin does in humans, yet millions of individuals use it as a sleep aid. In animals, it

Melatonin rhythms present an ideal example for understanding the basic concepts of circadian biology. Understanding melatonin rhythms will unlock many key concepts that help to explain other rhythms in the body.

can help measure the seasons, as we will see in chapter 6 when discussing seasonal affective disorder and mood. Surprisingly, bottles of melatonin are labeled not as a sleep aid but instead an antioxidant, yet this seems off target. Those who have taken melatonin know its ability to cause drowsiness, particularly as a naturally occurring substance.

Some parts of the physiology of melatonin have been carefully worked out. Intracellular molecular clocks control melatonin release in the pineal gland. Timekeeping signals from other brain regions, especially the SCN, which we will discuss later, control melatonin release. A schematic of this system is shown in figure 3.

While melatonin levels rise and fall at fairly predictable times, the total melatonin released varies widely. We do not know how this affects physiology. Could these differences be genetically determined? Could it be coding some information we still need to fully understand? Only time and research will answer this mystery.

Melatonin's Reading of the Sky

Melatonin has a striking rhythm. It rises and falls quicker than most other hormones, preferring to be at a high level or absent. In normal conditions, it rises shortly before bedtime and falls before we wake. This is partially due to

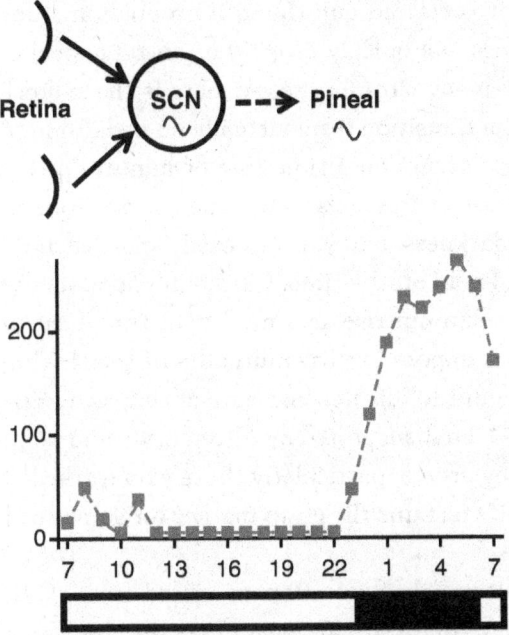

Figure 3 Top: The pathway from the eyes to melatonin production in the pineal gland. Circadian clocks in the SCN and pineal gland are noted. Bottom: A typical profile of melatonin production. *Source*: Simone Mäntele, Daniella T. Otway, Benita Middleton, et al., "Daily Rhythms of Plasma Melatonin, but Not Plasma Leptin or Leptin mRNA, Vary Between Lean, Obese and Type 2 Diabetic Men," *PLOS One* 7 (2012): e37123.

the fact that it is cleared quickly, so if production stops, melatonin levels will quickly drop. The central appeal of melatonin for many circadian researchers is these quick transitions. The transition from virtually no melatonin to peak melatonin occurs on a timescale of minutes rather than hours. Look at the melatonin time course from an individual in darkness, and you can easily spot when its rise occurs and read off the time. Circadian scientists use this time of melatonin rise as a marker of the phase of the circadian as opposed to the minimum of the rhythm described in figure 1. This new measure of *circadian phase* is called DLMO. First suggested by Alfred Lewy and popularized by many others, particularly those who worked at Harvard, DLMO remains the go-to marker for some circadian biologists.[2]

DLMO raises a key question in circadian biology: Should a rhythm time a specific event, here the rise of melatonin (e.g., as does the chimes of a grandfather clock), or provide a near analog output giving timekeeping information at any possible time (e.g., the hour hand of the grandfather clock)? On the one hand, the rise of melatonin can more specifically time an event, like the midnight chime of a clock. Yet if we take a single melatonin measurement, perhaps all we can determine is whether the body thinks it should be light or dark outside.

To predict whether it is light or dark outside, we need to know the timing of dawn and dusk. Work by Serge Daan in the 1970s, winner of the International Prize for Biology, postulated that the circadian system in rodents contains at least two clocks: one that tracks dawn and one that tracks dusk. The timing of dawn and dusk is hardwired into the specific circuitry in the brain, meaning the mechanisms of the timing of melatonin onset and offset depend on something beyond a single time measurement and include a prediction of the season. More commonly known as the evening/morning oscillator hypothesis, we will later delve into the specific neuronal circuitry within the SCN that performs this seasonal calculation.

We see a clear pattern where the melatonin rhythm tries to track the moving times of evening and morning. If it reflected these patterns exactly, the melatonin rhythm would not be interesting since it would not provide any additional information beyond what we see looking at the sky. If signals from the sky were obscured, such as by a heavy cloud cover or going into a cave, our timekeeping would automatically change, becoming out of sync with the correct time and the rest of the world. Thus the body needs a consistent signal reflecting its best guess of what is happening in the sky.

The seasons are a key part of understanding the sky. Melatonin's reading of the sky accurately predicts the seasons. In winter, melatonin rises earlier and is secreted longer than in summer. Again, though, this is not simply a reflection of light but instead a prediction of light levels, so this must be hardwired in the circadian circuitry. It can even predict the seasons differently when people live at different latitudes.

Light and Melatonin's Lunacy

Melatonin is extremely light sensitive, particularly to blue light. If the eye sees light while melatonin is produced, its production is quickly shut off. This is why laboratory studies of melatonin are done in near darkness. But again, this gets controversial quickly. Originally, melatonin was measured in dim room light (specifically 150 lux, where lux is the standard measurement of light's effects on the human visual system). This level of light is enough to read. Subsequent studies have shown that much smaller amounts of light can affect melatonin production. Light levels like 5 lux, much closer to what one would find in a darker night, are now being used. Recently this was further studied, and now we think even a fraction of 1 lux, a fraction of the light of a candle, could affect melatonin production.[3]

Nighttime light levels in the wild are high enough that they could be sensed by the circadian system and achieve much higher effects during a full moon. Given this, could our clocks be programmed to track moonlight and sunlight? There is strong evidence for this in animals.[4]

Evidence is emerging that moonlight does indeed impact our sleep. Could this be encoded through melatonin?

Our clocks have evolved through millions of years without worrying about the possibility of artificial light (except for the occasional fire). Candles started to change this until the invention of a practical light bulb transformed our lives. Now we can choose light levels to be whatever we want whenever we want. How should we then proceed? How does our ability to choose light levels affect melatonin production?

Kenneth Wright took his students camping in the Rockies during the summer without flashlights or exposure to modern lighting. Melatonin levels quickly reset to timing dawn and dusk when individuals were separated from modern lighting.[5] Wright then took his students camping in the winter. Melatonin production also expanded to the nocturnal winter night. The pineal gland releases melatonin when it thinks it should be night, which varies based on the seasons.

Taken together, this suggests that melatonin rhythms are the output of an intricate system within our body to predict external light levels. This system integrates signals over long periods to time and even guesses at the seasons. For me, the most fascinating part of this system is how it is responding to modern lighting in the real world in real time.

Other Factors Affecting Melatonin Production

Melatonin production is affected by posture, the state of the menstrual cycle, whether someone has smoked or taken aspirin, alcohol, or caffeine, and many other factors. Fatigue can affect how much melatonin is shifted by light. The timing of meals, however, does not affect melatonin production much.[6] As we discussed above, almost all human studies consider melatonin production in isolated conditions where social and environmental factors are removed. This is wise for a theoretical understanding of the circadian control of melatonin production. Yet in terms of a real-world understanding, many key behaviors might be missed. Is life "normal" without aspirin, caffeine, smoking, and alcohol, in the dark, and even when posture is regulated?

Melatonin is a specific biological signal integrating information from many different physiological systems. It is unique in terms of how closely it is regulated by light. But the real world is complicated, and many other factors likely play important roles.

Now that you understand melatonin dynamics, we can use melatonin rhythms to illustrate some key principles of biological rhythms that time events in the body. Since many features of melatonin rhythms are similar to those in other hormones or even rhythms in vital signs of the body, we are now ready to understand these crucial principles.

Clock Principle 1: Advances and Delays

Data on rhythms can be challenging to understand, and we will use melatonin rhythms as an example. Using circadian terminology, signals like light can either advance or delay the phase of a rhythm, as shown in in a *phase response curve*. Advancing the phase occurs when the clock ticks temporarily quicker than usual. This would bring us from, for example, US to European time. Light can also delay the phase of circadian clocks. If it does this, the light causes the clock to temporarily tick slower. When this happens, our clocks move from alignment with the continental United States toward, say, Hawaii time. Give light at one phase of the melatonin clock, light will advance it. Give light at another phase, and it will delay the clock. Light pushes the phase of the circadian system like I push my kid on a swing. As my kid approaches me on the swing, I push. This slows them down, delaying the rhythm. If I push as they move away from me, they go faster, and I advance the rhythm. The push is the same, just as the light input is the same. All that matters is when the signal is applied.

The effect of light on the circadian rhythm of melatonin production depends on what time the circadian system thinks it is, which depends on previous light measurements. If we knew the phase of a clock, we could predict how light would affect it. We often do not know the current state of the clock, though. To find this, we have to account for the timing of many light signals, accounting for what I did today, yesterday, and even last week. But it can be even more complex than that. Melatonin also phase shifts the circadian timekeeping of melatonin, so if an individual takes melatonin, that must be accounted for. Taking melatonin not only helps you fall asleep but also actually shifts the clock itself. Similarly, the effect of melatonin depends on the state of the clock. Taking melatonin at the wrong time will worsen jet lag. We need mathematical models to keep track of all of these signals.

Melatonin rhythms are advanced or delayed by light exposure. Light is sensed by the eye and transmitted to the SCN, two structures located at the bottom of the brain above the roof of the mouth, which keeps time (figure 3). When you first turn on a light, it can seem blindingly bright, but then your eyes adapt, and it does not seem as bright. A similar phenomenon happens with circadian signaling. The first part of the light signal is much better at changing the melatonin rhythm than the rest. Some evidence suggests that fast flashes of light (e.g., a fraction of a second in length) can phase shift the timing of melatonin.[7] The SCN then sends a signal to the pineal gland, which also has internal timekeepers that control melatonin secretion.

This has practical applications. Wake up in the middle of the night, and the first minute or two of light are most effective in signaling melatonin rhythms. After that, if you look away from the light and then look back at it, little

The effect of melatonin depends on the state of the clock. Taking melatonin at the wrong time will worsen jet lag.

phase shifting is lost. Many might think that just a quick burst of light would not have much effect, but it does. This is effective in the real world since such fluctuations are commonplace as we look closer and farther away from a light source. But remember to keep clear of even short, especially blue light at night or pay the consequences.

Clock Principle 2: Internal Period

A key parameter that can predict the behavior of a clock is its period. Think of an old mechanical watch. If it ticks too quickly or slowly, it will gradually lose or gain time, thereby giving less accurate predictions. A similar concept exists for other biological rhythms that keep time. Such rhythms have an "internal period," which tells us how quickly the rhythms would tick if isolated from all other external signals. This internal period depends on genetics and prior light exposure.

Clock Principle 3: Predicting Circadian Changes

Predicting advances and delays of a biological clock is much more complicated than a normal person can figure out. To better understand this system, tracking the underlying circadian system with mathematical modeling is helpful.

Researchers, including myself, have realized that mathematical models can incorporate many different datasets measuring the effects of external signals shifting melatonin. The models I first developed were published in the last millennium, and are still used in academia, industry, and the military.[8] Rather than focusing on model equations, we will concentrate on the structure of the model, which mimics the structure of the circadian timekeeping system.

To predict the behavior of this system, we use a simple model of light signaling in the eye that feeds into a simple model of an oscillator. We need to include a model of the eye since our eyes adapt to light signaling and this in turn affects timekeeping. Signals from the eye then signal to a simple model of an oscillator representing the state of the SCN. The oscillator equations are similar to those for the position of a pendulum or a kid on a swing. Yet the model's parameters are carefully chosen so that the model accurately predicts experimental data on melatonin rhythms. This is the basic model, on which much can be added.

Clock Principle 4: Interindividual Differences

Each person's biological clock is different. For example, a key parameter of the model is the intrinsic period of the human circadian clock. Just as a day is longer on Mars than on Earth, a day for one person might be longer than a day for another person unless corrected. Wearables can measure this, as we will see later. The reader might then wonder why it seems like we all live a twenty-four-hour day. Light, food, and other signals shift our clocks every day so that they match the twenty-four-hour world we live in. These intrinsic periods, however, do still affect when melatonin is produced. If you have a short *circadian period*, your melatonin will rise earlier, even if it rises at the same time each day. A longer intrinsic circadian period yields later melatonin secretion. So ideally, we would fit a model to each person's clock based on their wearable data.

Our work with permanent night shift workers suggests that shift work may change the state of the clock and working of the circadian clock. These workers constantly work the night shift and then try to live normally during their days off. We used models to predict the state of their circadian rhythm. They also were able to enter a lab at the Henry Ford health system to have their melatonin measured by researcher Phillip Cheng. When we compared the model predictions to the data, we found that the model consistently had more variability than the normal population. While we could typically predict melatonin onset to plus or minus an hour in normal individuals, it was double that in permanent night shift workers. So we must build a separate model for these night shift workers. Could the night shift permanently alter timekeeping in the circadian pacemaker? Would this be reversible?

Light, food, and other signals shift our clocks every day so that they match the twenty-four-hour world we live in.

Clock Principle 5: Crossing Time Zones

Humans can now cross time zones with planes so fast as to create time paradoxes, where the local time on landing is earlier than the local time before we left. Once we cross time zones, we create an incongruity for our circadian system, where its prediction of dawn and dusk is wrong. If only this could indeed reverse time. I would settle for just reversing aging.

Mistiming dawn and dusk may seem like an abstract problem, but to the millions of travelers who cross time zones, it is more than that. Melatonin onsets and offsets generally shift about one hour daily until they reach their correct times. Sometimes, though, they shift in the wrong direction. Correcting the phase can occur either by speeding up or slowing down time. So to go from New York to Paris time, we can advance by six or delay by eighteen hours. Advancing is much better, but sometimes people delay. Shifting to a new time zone (reentrainment) in the wrong direction is called *antidromic* reentrainment. But it can be even worse; noise and the variability of daily life can further confuse the circadian system, messing up time-keeping for even longer. Some of us experience this when traveling across the world and not feeling quite right for multiple weeks.

Melatonin rhythms can adjust to a new time zone much quicker if the correct timekeeping signals are given, such as if we see light at the right times of the day. These signals need to be carefully timed based on the current state of the circadian system and unintuitive (nonlinear) mathematical analysis. It took a good part of my career before a talented undergraduate (now a math faculty member at the University of Toronto) and I used mathematics to figure out when light should be given to adjust melatonin rhythms to a new time zone as quickly as possible.[9] Each of these light schedules depends on what time the system thinks it is and also which time zone we are adjusting to. Presented in an app, we gave this to hundreds of thousands of travelers.

People really want to open the app, click a button, and have no jet lag. But that sort of wishful thinking is just not possible. Nor would a simple suggestion, such as getting two hours of sunlight, work because the effects of light not only depend on when the light occurred but also on when your body thinks the light occurred, which is, by the definition of jet lag, off. So any proper algorithm to fight jet lag needs to figure out what time your body thinks it is, calculate how much and when the light should be given, and then present it to the user.

Getting people to follow schedules to shift their clocks may be even more challenging than figuring out what those schedules should be. Most of us care about jet lag, but only a few of us care deeply about it. Most of us care about jet lag enough to spend some time slightly modifying our lives for an hour or two a day over a week to avoid its ill effects. Some (e.g., professional athletes) care enough to spend even more time. No realistic jet lag schedule can be static. It must continually adjust to determine where the clock currently is. We will do other things, and few of us will follow any light schedule perfectly, so there will be much uncertainty for the researcher regarding how much each individual will follow a given schedule. It might even be better in some cases, particularly for short trips, to try to keep the melatonin rhythm set to the original time zone rather than adjusting it. A proper jet lag prescription resembles a GPS more than it does a simple bottle label. Perhaps this is why melatonin (which shifts the clock too) does not mention circadian rhythms or jet lag on its label.

Clock Principle 6: Rhythms Look for Specific Timekeeping Cues

Blue light shifts melatonin rhythms more than the light of any other color. This is convenient since the sky is blue on clear days. It suggests that the circadian system is tuned to sense light from the sky. Within the eye, cells called cones sense different colors of light. While each cone is tuned to sense a particular light color, no cone in humans can sense this blue light that the melatonin clock is particularly sensitive to. Exploring these properties of how light of different colors differentially phase shifts melatonin rhythms led to the remarkable discovery of a dedicated photoreceptive molecule in the eye, melanopsin, that senses blue light. Humans use melanopsin to measure light for the circadian control of melatonin. Melanopsin is not found in cones but rather in other cells dedicated to sensing light for the circadian system called intrinsically photosensitive retinal ganglion cells (ipRGCs).

If you were to change the properties of cones, you would see the world differently. For example, readers who are color-blind will see a painting at a museum differently than others. Changing ipRGCs or melanopsin would not change the colors you see in a painting. But changing these cells would affect our ability to organize our day with respect to the solar day. The opposite case also can be seen. Some "blind" individuals can lack functional cones but still have ipRGCs. They cannot "see" a painting, but their bodies unconsciously can organize their melatonin rhythms with respect to the solar day. Because of this, scientists call the melanopsin-ipRGC pathway subconscious vision.

As we described earlier, the timing of melatonin is most closely tied to that of dawn and dusk. The sky shows different colors at dawn and dusk than at other times. For example, the sky can turn beautiful shades of red, as mentioned in the age-old saying, going back thousands of years, "Red in the morning, sailors take warning; red at night, sailors' delight." There also is evidence that the circadian timekeeping system might use this information. Sensitivity to signals by dawn and dusk could be important in timing melatonin. If you do not want to interrupt your melatonin rhythms, you might want to try blue-light-blocking glasses. The circadian system is not wholly immune to nonblue light, but we can mitigate the effects of light at the wrong time of day by removing blue spectrum light. This is also why some devices change the background of their screens at night.

Clock Principle 7: Many Rhythms Are Often Present

Now is another time to fess up to the complexities of circadian timekeeping. I had to do just that at a scientific meeting. The biennial meeting of the Society for Research on Biological Rhythms is a sort of World Economic Forum for circadian rhythm researchers, except it is held in the United States rather than in Davos, Switzerland. There I gave what I thought was one of my best talks. Quite smug, I walked offstage to what I thought was an enthusiastic response and met Gisele Oda, a clock researcher and modeler from the University of Sao Paulo. First came the pleasantries, which went straight to my ego. But then came an insightful comment that stayed with me and caused much thought, like the pea at the bottom of the princess's bed. "Danny," she said, "but have you forgotten that there are multiple clocks?" Using the plural "clocks" rather than the singular "clock" complicates the problem. Some readers may have picked up on this earlier, where we spoke about *a single* circadian rhythm in melatonin, but then discussed timing dusk and dawn, two separate phenomena. Indeed, we will see separate rhythms in many other signals later in the book.

Melatonin onset and offset have different dynamics, probably because they beat to other clocks. Human data shows this, but beautiful experiments in rats done by neuroscientist Jimo Borjigin, building on original experiments by physiologist Helena Illnerová, show this better.[10] To get the best possible reading of melatonin, she put a probe into the rat's brain and measured melatonin continuously. Scientists can do this in a way that the rats recover from the probe-inserting surgery so that they seem quite normal in their activity. Still, this probe and the attached equipment can measure melatonin production in real time.

Jimo then had the rats follow different schedules, simulating shift work, jet lag, or other patterns that might shift the clock (there I go again, shift clock not clocks). What emerges is the most accurate picture of melatonin dynamics ever found. Melatonin onset and offset obey distinct dynamics. One might adjust quickly and the other much more slowly. Sometimes, one advances and one delays.

In humans, multiple melatonin peaks within a single day are possible (see figure 2). This sometimes occurs in shift work. Likewise, certain light schedules can stop melatonin production acutely and during future periods of darkness. These signals confuse the circadian timekeeping system, and it is tough to predict which phase of melatonin production will occur once it is restored.

Melatonin's Sleepy Message

During normal conditions, melatonin production is aligned correctly, starting before bedtime. This allows the body to prepare for night. Likewise, even if it is dark, melatonin shutting off signals the body that dawn is near and can aid the body in waking up. Even more helpful is that melatonin will reliably predict dawn and dusk if these signals are gone (e.g., someone spending time in a cave).

The timing of melatonin is complex, like many other rhythms in the body. Certain aspects of melatonin timekeeping are so well understood that mathematical equations can express them. But the modern world drives this ancient system far from its natural state, perhaps to chaos. Modern lighting can stop melatonin production at any time. One light pulse can also affect how the pineal gland will produce melatonin for many days. As everyday life allows pulses of light to occur at any time, this nocturnal signal may become less useful. With shift work, our nights look much like our days, and moreover, transcontinental travel allows us to have days or nights followed by nights. How important is this ancient "queen of the night" in the modern world? Are our bodies built to withstand the mistiming of this key signal? How much is this mistiming affecting our performance and sleep? These questions will haunt researchers for many years to come.

Humans will not return to living in the wild anytime soon. So as modern life seeks to disrupt our melatonin rhythms, how can we restore the natural rhythms of melatonin that our bodies were built for? We can take careful steps in this direction with a good app and blue-blocking glasses. Society can help with this too, from thinking about the color of streetlights for night drivers to indoor lighting in factories that produce products 24/7, or better design of hospitals to avoid sundowning, where night becomes mixed up with day for some patients. Studying melatonin rhythms also teaches us about how the body times circadian rhythms. Many examples follow in later chapters of how the body times other circadian rhythms related to the heart, sleep, and mood.

TEMPERATURE

Few numbers have such significant societal meaning as 98.6 degrees Fahrenheit (37 degrees Celsius). See it on a thermometer and you feel normal. Parents look worriedly at deviations from this number. Much above and we go to the doctor. During the COVID-19 pandemic, elevated temperatures caused people to be forced into isolation. There is so much importance placed on this number that one might think it is a physiological warning sign built into our DNA. But 98.6 is not hardwired in the body. The number comes from a large-scale study done in the 1800s. While it was innovative for its time, and having a specific number might be of clinical relevance, it is a foreign concept for our bodies, which evolved to live in nature. It is as alien as inserting a glass tube with mercury into our bodies.

Fever is a marker of infection, signaling concern to many body parts. Viruses, however, are not killed by the increased temperature alone. To sanitize and kill viruses that way would require much more elevated temperatures, which would cause more damage than it would help. Still, temperature can act as a signal for the immune system to ramp up its efforts. The higher temperatures allow the immune system to function at a heightened level, better fighting off infections. Yet when certain infections are present, temperature can go down.

Infection is just one of the many things encoded by temperature. Consider the example of my daughter coming home yesterday with a stomach bug. Nothing serious, but the mac and cheese she ate at school did not entirely agree with her. I will spare you the gory details of last night, but now it is just about time when school starts, and she is still in bed. Usually she would be enjoying a game she developed called "twenty rounds of attack daddy," perhaps as a delay tactic before the inevitable school starts.

I snuck into her room to take her temperature. It is above 98.6, but not enough to be called a "fever." But what does this mean? Typically her temperature would be significantly lower during sleep, so it is high considering that. As we will see later, though, this is a time of day when the temperature can naturally be elevated, so we would expect it to be somewhat high. Yet she is lying down, which can lower the temperature. Nevertheless, since she has a virus, we could expect the temperature to be higher, at least for some infections.

The body carefully generates many rhythms in temperature. These rhythms serve a host of functions. Not only do they prepare the body for activity during the day and help the body get to sleep but they also signal to clocks throughout the body what time it should be. Signals can reset the temperature rhythms, and rhythmic changes in temperature set us up for meals, activity, and of course, to fight infection. These rhythms persist even when external cues are removed. They are reset by meals in addition to light.

The center of this regulation is the *hypothalamus*, a basal part of the brain that sends off signals controlling many physiological processes. The SCN, the central circadian pacemaker in the brain discussed previously, exist in the hypothalamus. The hypothalamus also controls when we sweat to cool the body and signals muscles to create heat, increasing body temperature. While heat is generated separately from where it is controlled, the circulating blood carries heat throughout the body. This way, the hypothalamus acts as the thermostat, the muscles as the furnace, and the blood as the vents. Signaling molecules that mediate inflammation or infection are picked up by the hypothalamus. These signaling molecules, called cytokines, are the same ones that trigger sleep when sick. They can have a powerful effect on increasing temperature.

Diving deep into the temperature rhythms clarifies the body's physiological state, not just about infection, but a host of other physiological signals. Understanding your temperature rhythms can help determine ovulation time and other physiological signals. The rhythms are also beautiful in their function, such as in how some systems are engineered to resist some temperature changes but highly sensitive to others. This allows temperature rhythms to "keep the beat in the rising heat." Here we use temperature to show how complex regulating a vital sign is and how many factors need to be balanced. Many factors that need to be balanced in temperature regulations need to be balanced in other signs as well.

Temperature signaling can be complex. Sticking to a simple rule about a "normal" temperature seems fraught, and models could come to the rescue. The following sections should provide some information about how to interpret temperature. At the same time, none of it can replace a physician's advice. I am just a mathematician.

Daily Rhythms in Temperature

Traditionally, measuring temperature was the best choice if you wanted to measure a circadian rhythm within the body. Unlike melatonin, which requires collecting blood (or perhaps saliva) over a day, and then sending samples for costly and time-consuming biochemical analysis, temperature just needs a thermometer. Heart rate is another option; we will discuss that in the next chapter. While heart rate is just as easy to collect as temperature, it needs more modern mathematical techniques to understand because of the larger effect of activity on heart rate.

Measuring temperature rhythms goes back to Nathaniel Kleitman, a founder of sleep medicine, starting with an experiment in Mammoth Cave in 1938.[1] In hundreds, if not thousands, of subsequent experiments, what has been seen repeatedly is that if we keep all other environmental factors constant, there is an endogenous daily rhythm in temperature. Its endogenous period is also not exactly twenty-four hours, similar to the rhythms we looked at earlier. So if an individual is isolated from time cues for many days, the temperature rhythm will slowly drift out of phase with the strict twenty-four-hour day.

Temperature and melatonin circadian rhythms (which were the focus of chapter 3) are similar, and typically have a fixed-phase relationship. While melatonin rises just before bed and falls just before waking, temperature has its lowest value a couple hours before wake time. This makes sense since at night, the immune system usually rests, except if there is an infection, so the temperature signal kicks the immune system into a lower gear. Because of this, some have assumed that they measure the same "core" circadian pacemaker, and there is some truth to this. Both receive signals from the SCN.

Additionally, temperature rhythms respond differently to external signals than other rhythms (see figure 4).

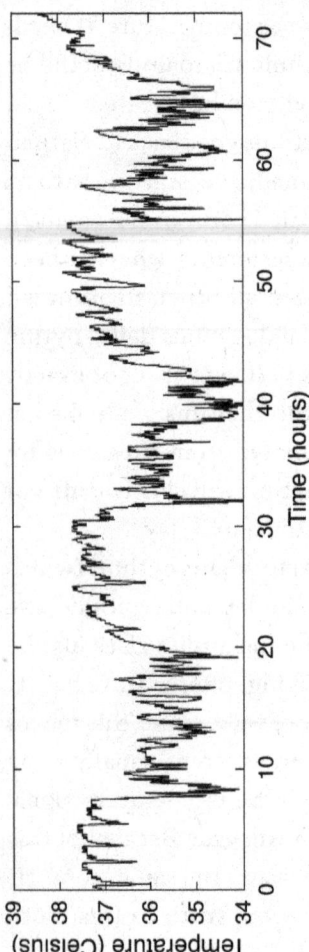

Figure 4 Skin temperature measured in a patient who has a fever at time seventy-two. *Source:* Christopher Flora, Jonathan Tyler, Caleb Mayer, et al., "High-Frequency Temperature Monitoring for Early Detection of Febrile Adverse Events in Patients with Cancer," *Cancer Cell* 39 (2021): 1167–1168.

Consider eating a high-carbohydrate meal at an unusual time, which surprisingly can affect temperature rhythms. Melatonin rhythms will ignore this and just follow the light information, as we explored in chapter 3. Temperature rhythms will shift, and the time of the minimum temperature will be different the next day and the following one because of the meal until other signals further change the rhythm.

The circadian rhythm in temperature can have an amplitude of up to 1 degree Celsius or more. What would be considered a fever at one time would not be a fever at other times of the day.

This becomes a fundamental question for immunocompromised individuals, as I have learned from oncologists Sung Choi and Muneesh Tewari. Certain cancer treatments leave their patients immunocompromised. For most of us, an infection that would simply pass in a day or two is a life-or-death scenario for their patients. So they are on a mission to better detect fevers in this population. Could wearables do this?

Sung and Muneesh collected wearable data from patients who might have fevers. Some of these patients did indeed have fevers. They shared their data with us to see if mathematical models of temperature rhythms could detect fevers with wearables better than traditional point-of-care methods, such as seeing if temperature was above a threshold.[2] We detected fevers hours before the current

The circadian rhythm in temperature can have an amplitude of up to 1 degree Celsius or more. What would be considered a fever at one time would not be a fever at other times of the day.

standard of care by looking for deviations from the rhythm rather than a fixed threshold (see figure 4). The hours just before a fever are critical in terms of coming up with the correct clinical course of treatment for the patient. More testing and optimization are being done by other groups, however, and this is needed before it becomes a widespread and routine part of clinical practice.

The Many Rhythms of Temperature

Francis Levi is the key innovator of chronomedicine (timing medicine to circadian rhythms). An oncologist in France, he developed the chronomedicine of chemotherapy, where chemotherapy is timed with respect to an individual's circadian clock. This makes treatments more effective, and giving chemotherapy at the right time can minimize side effects. Giving it at the wrong time makes us feel worse and is less effective. Levi along with collaborators Barbel Finkerstadt and David Rand at the University of Warwick measured circadian rhythms in temperature from wearables to determine when chemotherapy should be applied.[3]

Levi can measure temperature externally, such as with a temperature patch, or internally with a swallowed electronic pill. The pill will eventually pass through the subject, giving a day or two of data (in one case, I was surprised to hear that it stayed in the digestive system for over a week). What emerges is that these two rhythms, one of internal temperature and one of skin temperature, have opposite phases. Yet this matches physiology. As the body cools, it sends blood closer to the skin to lower the temperature. Skin temperature goes up, and body temperature goes down.

There are nevertheless many circadian rhythms in the hypothalamus, the part of the brain that controls temperature. The SCN in the hypothalamus consists of thousands of pacemaker cells. It sends many signals. Melatonin rhythms are timed by at least one timing dawn and one timing dusk, as discussed in chapter 3. There are signals to time sleep and napping, as we looked at in chapter 2. We have yet to learn what the multiple rhythms of temperature are. It gets even more complex because circadian clocks have been found in the hypothalamus outside the SCN. These other clocks regulate temperature as well.

For example, Kronauer and colleagues discovered a twelve-hour *hemicircadian* rhythm in temperature.[4] This rhythm is not regulated in the same way as the twenty-four-hour rhythm. The twenty-four-hour rhythm might shift in one direction and the twelve-hour one in another. One possible explanation for this data is that different parts of the temperature rhythm are regulated by different clocks, similar to melatonin rhythms, which may separate time at dawn and dusk rather than having a twelve- and twenty-four-hour clock. Future work will need to test these hypotheses.

Stopping the Clock

You may wonder how the circadian rhythm in temperature, which some refer to as a clock, might be similar to a grandfather clock, which you are much more familiar with. This temperature circadian rhythm is much more similar to a grandfather clock than you might expect. Surprisingly, after giving pulses of light at night to subjects in a study, researcher Megan Jewett observed that their circadian temperature rhythms disappeared. When the subjects reemerged several days later, their circadian rhythms in temperature reemerged with phases that were difficult to predict. Some were on Japanese time, and others on European time. The twelve-hour hemicircadian rhythms were much less affected by these light pulses, perhaps because a different noncircadian mechanism regulates them.[5] The behaviors that Jewett saw, similar to those seen in a grandfather clock, are intrinsic properties of almost all clocks. Clocks can be shut off and restarted at any possible phase.

This idea goes back to Arthur Winfree, an innovative interdisciplinary scientist who was a "genius," at least as determined by the MacArthur Foundation, and an inspiration to me. His book *The Geometry of Biological Time* explains this starting and stopping of rhythms with ideas from a branch of mathematics called topology.[6] It can be a bit hard to understand. Even Winfree called it "screwy results of circular logic."

Each signal a clock receives affects its amplitude and phase, as seen in chapter 1. So far, we have focused on the phase of rhythms, so some discussion about amplitude is helpful. Sometimes we would like to increase the amplitude of an oscillation. For example, if I am at the playground with my kids, they want me to increase the amplitude of their oscillation on the swing. A kid on a swing is a reasonably reliable oscillator—enough so that it can be modeled accurately by simple Newtonian physics. Pushing kids at the proper phases can achieve a larger amplitude. Pushing them at other phases of their swing cycle can lower the amplitude of their swing, which I do when we need to go home. Eventually I can push them to stop at the lowest point in the swing. Although the swing is designed to oscillate, it sits motionless at its lowest vertical position. When the next adult and child arrives, they can start the oscillation at any possible time, setting up any new phase relationship between the local time and the oscillation. Similar behaviors can be seen with the pendulum of a grandfather clock.

Jewett discovered a similar phenomenon with human temperature rhythms (a.k.a. the swing) and light (a.k.a. the push). As light pushes temperature rhythms to new phases, transiently speeding or slowing its rhythm, it also can increase or decrease the amplitude of the rhythm, depending on when it is received. Just like me pushing my kid to stop the rhythm, ceasing it at its lowest vertical point, Jewett pushed the temperature rhythm to a state where it stopped too. She happened on this state unexpectedly, but it was just as real as the steady state on the swing. Just like the swing can be restarted at any time, the temperature rhythm can be restarted as well.

There is one difference between many biological clocks and a swing or pendulum. If we do not push the swing, eventually, by itself, the rhythms will wind down. The swing needs us to push it constantly. Circadian rhythms in temperature or other biological processes, though, are self-sustained. They might use adenosine triphosphate or other energy sources, but they have intrinsic rhythmicity, just like the springs of a watch. But even if a clock is wound, if the pendulum stops at its lowest point, it may not swing until it is pushed from its lowest point. Similarly, when the temperature rhythm stopped, it did not quickly reemerge.

Homeostatic Regulation of Temperature

In addition to the circadian rhythm in temperature and increases in temperature due to infection, the hypothalamus homeostatically regulates temperature, similar to the function of your thermostat. So circadian clocks and other signals, such as cytokines indicating infection, set the target temperature. During the day, this temperature set point increases, and even more so if there is an infection. Yet more important than the set point is how temperature is generally regulated. Life has lots of factors that change body temperature, so there is no one normal for temperature rhythms in the real world. We can step in a sauna or hop in a freezing lake (some people find it pleasurable to do both). We can exercise or just sit still in a cold room. Kids sometimes wear too many layers (or more often, not enough). This represents a key signal in temperature rhythms.

As temperature increases beyond the set point, the hypothalamus triggers the body to decrease its temperature by sending blood closer to the skin or sweat. When more heat is needed, muscles or adipose tissue (fat) are recruited to do the job. The timescale for all of this is about an hour. Temperatures go up and down slowly in the body.

Emery Brown gave a crucial insight into how to track this.[7] His genius is also well-documented, and rightly so (he is one of only a handful of scientists to be elected to all three of the National Academies of Science, Engineering, and Medicine, sort of an academic greatest of all time). Yet this early work on temperature is still underappreciated. The key mathematical difficulty is that this homeostatic regulation interferes with our measurements of other processes. How can we tell the difference between temperature falling because of infection or another reason the hypothalamus is trying to counteract? Emery figured out the statistical methods that can mathematically disentangle these multiple signals. We will return to these methods in more detail when we discuss heart rate (in the next chapter).

Meals, Posture, Activity, and So On

Temperature also reflects the basic workings of the body. Some activities generate heat, and this will be seen in temperature rhythms. For example, it takes a fair bit of work to digest food. Muscles contract in the stomach and intestines. All of this creates heat. As we walk and run, heat is created and dispelled. Even posture has consequences on heat since different muscles are at work and blood flow is different. All of these effects can be seen in temperature. Moreover, it means that studying individuals in the ICU, where there is little motion, posture is fixed, and there are no standard meals, yields temperature time courses where the effects of circadian regulation and cytokines as well as other endogenous factors can be more easily seen.

But even in the real world, outside the ICU, we can use ML to extract these different signals, especially with the help of multiple data streams. If steps are recorded, we can determine how much temperature should be increased and account for this along with homeostatic effects. Meal timing might be detected from temperature rhythms (as well as heart rate, as we will see). Individuals could also annotate this to help the algorithms.

One additional important factor we will only discuss in passing is the effects of drugs. We know these can affect temperature. For example, if my daughter shows a temperature, I will give her ibuprofen to reduce the fever. Likewise, caffeine can decrease the heat lost through the skin by constricting blood vessels. Such real-world effects are important to consider.

What remains more complex is the interindividual differences. There is wide variation between the rhythms we see in one individual versus another. Some of this variation has to do with genetics. For instance, genetics can determine whether our intrinsic circadian clocks tick faster or slower. Some of it might have to do with prior history. The previous alignment of the many clocks in the body, say, might cause larger amplitude circadian rhythms. Also, depending on a person's mass, fat-to-muscle ratios, and so on, the temperature's homeostatic properties might differ. For these reasons, the analysis of temperature time courses needs to be personalized. This is slowly becoming a reality, yet large-scale datasets from many people in many conditions are needed to be accurate.

Temperature as a Timekeeping Cue

One of the most remarkable properties of circadian time-keeping is that every possible cell in the body can keep time. Daily timekeeping is genetically hardwired through special genes and proteins that interact in a network structure that mathematically can produce oscillations. Each of these clocks in individual cells can time events, but they all need to be synchronized or chaos will ensue. The common analogy in the circadian field is a symphony of clocks, where the conductor is the SCN. It is now widely accepted, however, that individual cells often ignore or only partially integrate signals from the SCN. But what is the timekeeping signal from the SCN? One possible explanation is temperature rhythms.

Ethan Buhr and his mentor, Joseph Takahashi, hypothesized that the temperature rhythm is used to entrain circadian clocks throughout the body, perhaps based on earlier work by Steve Brown and Ueli Schibler.[8] The key finding was that when tissues of mice's bodies were removed and grown in an environment with a temperature cycle mimicking the animal's natural circadian rhythm in temperature, the clocks in tissues such as the lung or pituitary gland entrained and matched the phase of this imposed temperature cycle. Amazingly, they found that the SCN ticks without regard to this temperature cycle. This was not a property of individual cells of the SCN. When they stopped the coupling between individual cells in the SCN or the SCN was cut into pieces, the individual cells or fragments of the SCN followed the temperature cycle. Something about how multiple clocks talk to each other within the SCN supplied its resistance to temperature cycles.

Thus temperature rhythms signal beyond fevers, telling the body about circadian rhythms. Temperature rhythms controlled by the SCN synchronize rhythms in the rest of the body. Yet cells in the periphery might also feel external temperature patterns more closely than those in the hypothalamus. Moreover, they react to this non-SCN information. So the oscillators hear the conductor (SCN) but can choose not to follow it. They likely react to other temperature signals as well, such as temperature increases during infection. Their clocks will sense other temperature signals and arrange their phase accordingly. Meaning that this timekeeping signal through temperature is not absolute.

Rhythms of Sleep and Body Temperature

But what if we lived a non-twenty-four-hour day? Let us return to Wright's experiments, having individuals follow a non-twenty-four-hour day, as described in chapter 3.

What will happen to temperature rhythms? Temperature rhythms will "free run," as we discussed for melatonin, and have their circadian rhythms return to an approximately twenty-four-hour rhythm. The temperature signal will have this rhythm and the induced non-twenty-four-hour sleep-wake rhythm, whereby body temperature decreases when we are asleep.[9] Some days, these will be in phase; on others, they will not.

Looking at the raw temperature reading requires some analytic skills since two rhythms will emerge (along with other possible effects). One will follow the sleep-wake cycle simply because our body temperature falls whenever we are asleep. The other will follow the intrinsic circadian rhythm, which may not be aligned with the sleep-wake cycle. Separating these rhythms can be done mathematically, especially if their periods are significantly different. A key question, though, is what individuals would feel when this happens, not just immediately, but over an extended period. The health consequences of this "forced desynchrony" are beginning to be understood. Our bodies are just starting to figure out how to adapt to this as modern shift work, with modern lighting, has only appeared in the tiniest fraction of evolutionary time and will not be solved by evolution any time shortly.

Cliff Saper and colleagues, in experiments with rats, provided evidence for what happens in this case. His group showed that neurons from the SCN send signals to two separate hypothalamus regions to control sleep-wake and temperature rhythms. Block one pathway and you can find attenuated circadian rhythms in sleep but not temperature.[10] Block another pathway and you will find attenuated circadian rhythms in sleep but not regular temperature rhythm. These are also separate from pathways outside the SCN that control fever response. Thus the SCN can control sleep and temperature rhythms separately. Just because the signal to sleep at certain times is likely being overridden by the homeostatic control of sleep does not mean that the temperature signal is also being overridden.

Temperature Rhythms Predicting Ovulation

There are other important rhythms in temperature. Understanding the effects of circadian timekeeping, sleep, fevers, activity, and meals is part of the story, as I learned the hard way. This brings me to the most difficult interview I have ever done. Shortly after one of my scientific papers was published, it received media attention, including the opportunity to go on a live radio broadcast. I was particularly excited about this and drove to the studio while practicing what I would say. I waited outside the room as the host finished her previous work. Then the "Live" sign went dark. She quickly greeted me, set up the microphone just by my mouth, and "Live" was lit up again. In my mind, the ensuing monologue went something like, "I am excited that Dr. Forger, a world expert on biological clocks, is joining us today. I have been telling women to lean in closely and not to miss this." Uh oh. "Dr. Forger, I am excited to have so many women listening in. Our biological clocks are ticking much faster than I ever thought. Many older women want to get pregnant. Please, Dr. Forger, tell us how to turn back time." Silence. This was not the kind of biological rhythms we explored in our paper.

I did not know it then, but temperature rhythms can be helpful in this regard. A fascinating and useful application of monitoring circadian rhythms in temperature is to predict fertility. To become pregnant, women must release one (or multiple) eggs. These come from follicles. Of the many possible eggs, just a few (typically one) get chosen for growth during each cycle. During a period when estrogen levels peak, a woman is fertile. After this period, ovulation occurs, and the estrogen is replaced by progesterone.

Estrogen and progesterone, also under at least partial control of the hypothalamus, strongly affect the daily circadian temperature cycle. These impact the amplitude, phase, and most important, mean temperature during the rhythm. Just before ovulation, the temperature rhythms achieve their lowest readings when estrogen levels are high. Just afterward, when progesterone levels peak, the temperature rhythms are highest. This is how wearables predict ovulation cycle. These differences can be a degree Fahrenheit or more. Hence the monthly rhythm in temperature is a crucial predictor of fertility. ML approaches can use this rhythm to inform couples about fertility better too.

But again, much more work needs to be done. Prior research on temperature rhythms has disproportionately focused on men, which needs to be corrected. We need large datasets to understand how temperature signaling of the state of the menstrual cycle interacts with other real-world stimuli. If more real-world information was available, this topic could become another chapter in and of itself.

Temperature Compensation

We have discussed many factors that could increase or decrease the temperature of cells within our body. Cells also time events with intracellular biochemical clocks, and each reaction in these clocks is biochemical. Would these increasing reaction rates throw off circadian timekeeping?

Let's do a quick calculation. Most biochemical systems increase their rate two- to fourfold (we will take the lower value now to be conservative) when temperature rises by ten degrees Celsius. So consider a one-degree increase, and your circadian clocks should increase their speed by

10 percent. This means the clocks are now living an approximately twenty-two-hour day (perhaps even shorter if we are not being as conservative). Such clocks lose two hours every day, which is worse than useless.

Colin Pittindrigh discovered the answer to how clocks function at different temperatures in an outhouse high in the Rocky Mountains in the 1950s. A father of circadian biology, Pittindrigh took two pressure cookers and filled them with flies to study the insects' rhythms.[11] He put one pressure cooker in a cool stream, thereby reducing its temperature. He put the other in an abandoned outhouse, which got extremely hot in the summer sun at high altitudes. Although the changes in temperature initially affected the rhythms, he observed that both cases returned to a twenty-four-hour rhythm. Therefore rhythms are not independent of temperature but instead can compensate for a long-term temperature change. This allows clocks to be robust to overall changes in temperature and yet sensitive to temperature signals that could confer timekeeping information. Pretty cool.

How this happens has perplexed scientists for the better part of a century. My best guess comes from work we did with David Virshup at Duke's medical campus in Singapore with the National University.[12] We found that the circadian protein PERIOD2 switches from a more or less stable state, and this switch is sensitive to temperature. The more stable state slows the clock at higher temperatures, balancing the effects of increased temperature on other reactions. There are likely other parts of this story that will fascinate scientists for decades.

Temperature as Mercury, the Winged Messenger

Composer Gustav Holst's musical description of Mercury taking messages between the celestial beings is a great metaphor for the circadian rhythm in temperature. All cells naturally are affected by temperature. Reactions within the cell go faster as temperature is increased. Special signals are produced as temperature changes. Thus it makes for a handy signal in the body.

Measure temperature with a wearable, and you can listen to this signal. You can see what time your hypothalamus, particularly your SCN, thinks it is. But it also indicates if you are fertile, have an infection, took ibuprofen, went running, or even the composition of fat to muscle or your size. How it sends all of these signals with just one line is quite remarkable. The body listens to temperature except for the SCN, which sets it. As temperature becomes more ubiquitous in wearables, we will learn more about the current state of our bodies, not just when to see the doctor, but perhaps how well synchronized our bodies are, when to do chemotherapy, and even when to stop or start exercise.

Measure temperature with a wearable, and it indicates if you are fertile, have an infection, took ibuprofen, went running, or even the composition of fat to muscle or your size.

Our temperature rhythms are uniquely tailored to each of us. Some individuals have temperature rhythms with larger or smaller amplitudes. Some have higher or lower average circadian temperatures. Genetics can change whether your temperature rhythms are early or late. Sometimes temperature rhythms can be shut off or even show multiple superimposed rhythms. There is no normal temperature, just as there is no normal world. We just need to decipher the message.

Mathematical analysis allows us to understand the language of temperature rhythms in the body. ML can interpret and separate these signals so we can tell whether the increased temperature is due to the circadian clock, a homeostatic response, or a possible infection. We can combine this with mathematical models tailored to your prior data. These signals can then issue warnings that you might have an infection or should get out of the sun while running, or if it might be a good time to be romantic.

HEART RATE

As I walked into my PhD adviser's office near Washington Square in New York City, in fine print, his doorplate read, "Charles Peskin, . . . Secrets of the Heart Revealed." Though he started medical school, he decided to switch to mathematics after finishing a PhD because he saw the importance of new technology (computing supercharged with mathematics) in understanding physiology. Charlie has worked on everything from hearing, blood flow, circadian rhythms, and mRNA, and for this work, he was classified a genius by the MacArthur Foundation. His heart, though, was always in . . . wait for it . . . the heart, and some of his infectious enthusiasm for the heart must have rubbed off on me.

As we will see, there are many rhythms in heart rate and many reasons to be fascinated by it. Like melatonin or temperature, heart rate signals many aspects of the state of the body. We aim to understand these "secrets of the

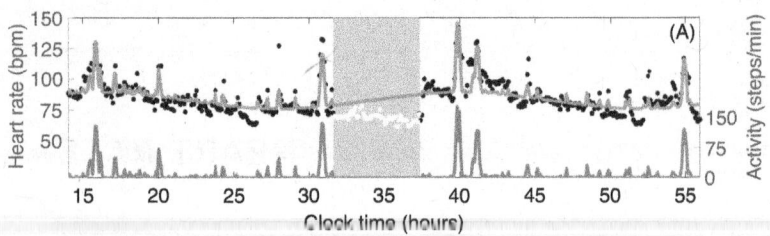

Figure 5 Heart rate measurements (dots) from a wearable. The sleep period is shaded. The bottom trace is the individual's activity. The upper curve predicts heart rate when the factors of our framework are accounted for. *Source:* Clark Bowman, Yitong Huang, Olivia J. Walch, et al., "Characterizing the Human Heart Rate Circadian Pacemaker through Widely Available Wearable Devices," *Cell Reports Methods* 1 (2021): 100058.

heart," which will differ from melatonin and temperature signals. Here we show how a particular framework for decomposing signals in heart rate can yield much information about a person's physiology. Similar frameworks can be applied to other signals as well, but they are best described in terms of heart rate. When a parameter is central to the framework, we will indicate it in italics. Moreover, you have probably heard of some of these terms, so we provide a holistic description.[1] To demonstrate the importance of this framework, we end by showing how the framework can be used to detect COVID or other infections before an individual exhibits overt symptoms.

Before discussing the details of the framework, illustrated in figure 5, it is helpful to understand more about the data. Millions of devices, such as the Apple Watch or

Here we show how a particular framework for decomposing signals in heart rate can yield much information about a person's physiology. Similar frameworks can be applied to other signals as well.

Fitbit, are currently collecting heart rate. You may have wondered what all of these measurements can tell us about our physiology. We have studied this in thousands of individuals with a large dataset known as the Intern Health Study among others. Over a year, thousands of first-year MDs (interns) have shared data, which has been recorded by these wearables as the doctors work during the day and night shifts. Srijan Sen, the leader of this study, has also collected genomic and daily mood scores along with other data types. After spending years pondering this data, we see clear patterns in heart rate rhythms, which we will now describe.

Basal Heart Rate

The most common measure of heart rate reported is *basal heart rate (BHR)*. There are several ways to measure this, but the basic idea is to measure heart rate when activity does not occur and other factors that affect heart rate are controlled. Devices can then measure BHR. What is reported by wearable companies, however, often needs to account for more factors. For example, heart rate has a circadian rhythm, so "basal" heart rate could vary based on what time of day it is measured.

BHR is considered a measure of overall health. Typically, for healthy individuals, it will be sixty BPM or less. A well-trained athlete can have a BHR of much less, perhaps even forty BPM. Unhealthy individuals might have BHRs significantly higher than sixty BPM.

Breathing Rate

Breathing rate can be inferred from heart rate, or more specifically, we affect heart rate as we breathe. Slight but perceptible increases in heart rate can be seen as someone takes breaths in and a slight slowing of heart rate as we breathe out. So one can look to see how these rhythms interact over an interval or a minute or more. In theory, breathing rate should be accounted for when measuring other factors in heart rate. Yet we look at timescales of minutes or longer, where breathing rate will have little effect on average.

Stress and Shock

In high school, I tried to give blood. Someday I might need some, so I thought this was only fair. Plus several of the "cool" people were doing it. But my experience was far from cool. I suddenly felt faint. My heart rate surged, and I started to sweat. People began to crowd around me, and my experiment in giving blood was soon over. I had a vasovagal response—something that is not that worrisome and perhaps even psychosomatic in some instances but enough for the nurse to tell me to avoid giving blood in the future.

This is one example of stressful events that affect heart rate. We have all had similar experiences. The high school student sees an email from a college they applied to. The author sees an email from a publisher considering their work. The performer stands on stage for the first time to face a live audience. It is a typical response, and the increase in heart rate is likely due to *adrenaline*. Sometimes we can almost taste the adrenaline. Adrenaline also plays other roles in regulating heart rate, rather than these hopefully rare events. More about that later.

Meals

Heart rate is determined by the effects of hormones, particularly adrenaline. Adrenaline increases heart rate and is a general indicator of stress in the body, as we discussed earlier. One might hear of adrenaline junkies, people who work high-pressure jobs or engage in risky activities for the boost they feel. It does not need to be just athletes; think of characters in the old TV show *The West Wing*. It is not only activity or stress that can trigger adrenaline. Meals trigger adrenaline too. This makes sense since, as we explored earlier, eating (or more precisely, digesting) takes more energy than we typically imagine it would. In response to a large meal, the body releases adrenaline, increasing heart rate. As the blood pumps faster, it provides extra nutrients to support digestion. The process takes about the right timescale until the food is passed through the stomach and into the intestine.

Unfortunately, we cannot directly measure adrenaline with a wearable yet, although some promising technologies are developing.[2] We can nevertheless see the effects of adrenaline on heart rate wearable data (see below). Staring through hundreds of daily heart rate measurements, a pattern emerges. Three small peaks in heart rate typically appear each day, decreasing with the same timescale one would expect from adrenaline: one shortly after waking, one around noon, and another in the early evening. These are likely due to daily meals and the adrenaline secreted in response.

But these are not the only adrenaline spikes we see. Other activities can affect adrenaline, and long-term standing can affect it. More data is needed to show how adrenaline affects wearable data and what types of activities cause adrenaline spikes in heart rate. Acute stress triggers adrenaline. Could it help detect chronic stress or affective disorders? These could be detected from baseline adrenaline levels, which could affect BHR.

Activity

Heart rate, how fast the heart beats, determines how quickly nutrients and oxygen are provided to the body. We all know that heart rate increases during exercise. As muscles consume energy, they need more oxygen, which is transported in the blood by hemoglobin. Blood contains many other nutrients, signaling molecules, and even infection-fighting cells. The heart carefully considers these needs, adjusting its heart rate to meet metabolic and other demands.

Walking expends energy. As you walk, you would expect your heart rate to increase. In general, this is true. But take a couple steps and then stop. Your heart rate does not automatically increase. Some amount of activity can occur with the BHR. When more activity is needed, heart rate goes up. Still, heart rate cannot increase further past a certain amount of activity. Heart rate indicates, to some extent, the level of physical exertion a person expends.

The increase in heart rate based on activity is personalized. Someone in great health, such as someone who runs marathons, would not see that much of an increase in heart rate as they start to walk. Now consider a person living a sedentary lifestyle. Taking a couple steps yields a dramatically different response, with a large increase in heart rate. All of this depends on metabolism, how efficiently we expend energy, how well our muscles work, what we weigh, and other factors. For this reason, we need to understand how our heart rate changes with exercise in a personalized way.

The personalized parameter we developed to measure this is *heart rate increase per step* (HRpS). To get this, we take data over a day and look statistically at how much heart rate increases on average for each step taken in a minute. This is an average measurement; as mentioned above, the first couple steps might not increase heart rate much, which we can account for. HRpS will vary based on an individual's overall health, with a lower HRpS generally indicating more excellent health, and also on a particular day, such as whether we have an infection, as we will see later in this chapter.

HRpS also has a clinical counterpart, although a gold standard validation has yet to be performed. The six-minute walk is a commonly used test to determine cardiovascular health. Individuals are asked to walk for six minutes, and heart rate increases are recorded. By doing this for six minutes, the test measures not just initial transients but the effects of steady-state activity too.

There is a balance between BHR and HRpS mainly due to a maximal heart rate that can be attained. The heart can only beat so fast; you might experience this if you push yourself athletically. The range between this BHR and the maximal heart rate is determined by the number of steps you can take per minute at your maximal heart rate and the HRpS parameter. In theory, the lower the BHR and smaller the HRpS parameter, the faster you can run.

One technicality needs to be explained when calculating this parameter, at least with the data reported by most wearables. Wearable data is frequently "binned," mainly to save storage space, so we do not get data on each heartbeat, but a heart rate averaged over several (typically, for example, five) minutes unless the wearable is in a special high data collection mode. But imagine you start to run in the last minute of a five-minute bin and then stop. Heart rate does not instantaneously increase since the BHR can support the first couple steps, so we will likely record more steps in that bin and yet no increase in heart rate. Now more important, consider the next five-minute bin. Since the running has ended, the number of steps is low (perhaps zero), but the heart rate is now elevated due to the running from the last bin. The net effect of all of this is to increase the heart rate in the later bin, where fewer steps were taken, and decrease the heart rate in the earlier bin, where extra steps were taken. The net effect of this data binning is to decrease HRpS and bias its calculation—something that we can take into effect.[3]

Posture

If you lie down, blood needs to be circulated horizontally. When standing, your heart needs to pump blood against the blood's weight and gravity's effects. You have probably experienced the effects of this. Sometimes (luckily rarely), I wake up in the middle of the night, stand up to walk to the bathroom, and suddenly feel faint; something is wrong, and my heart starts pounding. What has happened?

My heart rate at the beginning of this episode is low for several reasons. Heart rate is low during sleep (see below). My *circadian rhythm in heart rate* (CRHR; more about that later) might also favor a lower heart rate at this time. If my posture is supine, my heart rate is lower as well. Suddenly I stand, and blood needs to travel to my head. My heart rate is too low, and my heart is not working hard enough. This makes me feel faint. But the heart senses all of this and quickly increases its heart rate. The technical term is postural orthostatic tachycardia syndrome (POTS), which can be serious for some individuals.

We calculate HRpS during the waking day, so we can assume, without too much error, that individuals are not lying down. They could be sitting down, and when they do, their heart rate is less than when walking. Heart rate is increased by both standing up and walking. You might also ask about standing and not walking, but again, people tend to only stand a little without walking, so we do not need to consider this here (but can in detailed modeling). Thus taking steps is a decent proxy for standing up. As people start walking from sitting, HRpS increases more than it normally would since it reflects the combined increase in heart rate from changing posture and increasing activity. As we walk for longer, however, we measure only the effects of changing walking speed on heart rate and not posture. As we walk longer, causing more steps to be recorded, HRpS is lower since it does not reflect postural effects. This can be taken into account to give a more accurate measurement of HRpS.

Correlation on Two Timescales

Here is an experiment: Run as fast as possible with a wearable heart rate monitor for five minutes and then stop. Watch what happens. You should see a steady decline in heart rate after stopping. After a minute, you should see your heart rate drop significantly, halfway or more to your BHR. After a couple minutes, you are back to normal. This shows that your heart rate is "correlated," (*heart rate correlation*) meaning that its current value depends on the activity at previous times. We must account for the binning we examined earlier. Does it need to be accounted for in the same way as we did for correlations in temperature?

Much of this correlation is lost in the datasets we can access since the data is binned and has too many steps combined to see these effects. It is hard to measure since the effect lasts only a minute or two, and we only have data every five or ten minutes. Still, I tried and asked the computer to find correlations in the heart rates we had measured. It consistently found a correlation across many datasets as well as on a much longer timescale. This timescale was close to the correlation in temperature data, about one hour, and we call it the *correlation time constant*. What could be going on? Adrenaline cleared from the body at this timescale. It is likely that the correlations the computer often found were due to adrenaline, and we are measuring its effects on heart rate.

We can take this analysis a step further by looking specifically at the one-hour timescale and for triggers that cause sparks in adrenaline. A stressful event may have occurred. Perhaps some physical activity triggered adrenaline. We describe these effects as the *Internal Heart Rate Variation*. We call it noise since it is difficult, if not impossible, to predict in advance. It is internal noise that distinguishes it from measurement noise, which depends on the accuracy of the device we are using, which with modern devices, can be relatively low. Errors from wearables (measurement noise) tend only to affect the current measurement, and a different measurement and error will happen the next time. Internal noise comes about from responses from the heart rate regulation system (e.g., hormonal) that are long-lasting and persist for several (or perhaps many) measurements. If we better understood adrenaline dynamics, we could predict these changes and then not call them noisy.

Circadian Rhythm

Kleitman's experiments with people living underground noted a circadian rhythm in heart rate. One concern Kleitman raised, though, was that the rhythm in the sleep-wake (activity) cycle could mask the rhythm, just as we discussed for the circadian rhythm in temperature.[4] Many future studies have shown that the rhythm persists when an individual stays awake and supine, and even when regular meals are eliminated. This rhythm is quite strong, raising and lowering heart rate by several BPM.

Yet there is one wrinkle. A fascinating sleep study had individuals sleep on a non-twenty-four-hour cycle. As they did this, they slept at all possible circadian cycles. Piecing together the data from many days, and separating the data when individuals were asleep and awake, a CRHR could be determined separately while an individual was awake and asleep.[5] Amazingly, these rhythms had different phases. This physiology is reasonable and perhaps even expected if one thinks about it. Different systems govern heart rate during sleep and wake. Because of this, we typically do not include data during sleep when calculating the CRHR.

This still leaves about sixteen hours of data to explore—enough mathematically (but barely) to determine the CRHR. Some individuals only use their wearables sometimes, though. We found that as individuals wore the device less, such as only twelve or eight hours during the day, the CRHR became more challenging to determine.[6] This has to do with the mathematical properties of oscillations. Luckily, some individuals keep their wearables on for most of the day.

The CRHR has many properties shared with other circadian rhythms in the body. It typically peaks in the afternoon and falls at night. This allows the body to be ready for activity during the day and prepares the body for sleep at night. Rather than just following an activity or sleeping, however, the CRHR predicts it. Since we had thousands of individuals' activity patterns and heart rates, we measured how activity shifts the CRHR, not immediately, but how activity on one day would affect the CRHR on subsequent days. This might seem counterintuitive. Why would doing an activity at a particular time one day affect the CRHR on subsequent days? The answer is that engaging in activity at a particular time on a specific day shifts the CRHR on future days so the body anticipates activity at that time on future days.

You may have experienced this previously. One day, wake up in the middle of the night and do a lot of activities, such as going for a run. The next night around that time, your heart rate may be higher, and your body is ready in case you decide to have more activity. This likely is an ancient response evolutionarily. If a cave dweller woke up and had a lot of activity in the wild at night, it could easily indicate that it was a peak activity for a predator or food

The answer is that engaging in activity at a particular time on a specific day shifts the CRHR on future days so the body anticipates activity at that time on future days.

source. So it would make sense to be ready the next day, mainly if that activity occurred regularly.

Yet each individual's heart rate is different, as is their CRHR. Other individuals show different amplitudes and phases of this rhythm. Many interns in Sen's study have at least a year or more of wearable data. This allows us mathematically to ask, personalized to just one individual, how activity at different times of day shifts the CRHR. We found a consistent pattern among individuals for how activity at particular circadian phases advances or delays the CRHR on future days. The *circadian amplitude* nevertheless varies greatly among individuals, as does the sensitivity of their CRHR to activity. For example, the effects of activity on my CRHR were quite different from those of my postdoc Clark Bowman's activity on his CRHR. This is likely due to the differences in our activity levels. Individuals who typically have lots of activity (e.g., ten thousand steps or more typically a day) should have less of an effect of each step on their CRHR.

Anticipation of activity in heart rate differs from what we saw for melatonin, which anticipates light (or more precisely, darkness). So how does this heart rate clock work? It turns out that cells within the sinoatrial node, the part of the heart that sets heart rate, have internal clocks within them. As activity changes, these cells will increase or decrease heart rate, and their intracellular molecular clock listens to these signals to guess when activity will occur on the next day. These internal clocks also listen to the effects of light. Tim Brown's group at the University of Manchester found a direct pathway from the cells within the SCN that sense light to the pacemaker cells in the heart. This allows for light to increase heart rate too.

The Intern Health Study dataset is particularly rich to explore CRHR since interns are shift workers. They often work night shifts and then switch back to day shifts. Two patterns in their data are worth noting. First, the experience of different interns varies quite a bit. For some, the CRHR ignores the shifts. For others, the CRHR adjusts, or at least tries to do so, for each shift. This could also be due to the different responses the CRHR and other circadian rhythms show for different shift workers. Additionally, we see that shift workers never fully adjust their CRHR to the new shift. Some partially adjust their CRHR, shifting it a couple hours, which allows them to sleep a little better than if the adjustment had not happened.

Our work also predicts that the CRHR can go out of phase with the rhythm of melatonin. An example of this was during the initial lockdowns of the COVID pandemic.[7] Shortly afterward, we released an app and asked people to send us their wearable data and the date when they began physical distancing. This physical distancing changed circadian timekeeping cues. Individuals mainly stayed indoors. This meant much less outdoor light. Their meal and work schedules changed with remote work. Different individuals moved in together, and for reasons we do not fully understand, their CRHRs became more aligned. It might be because they synchronized their activity, meals, or even sleep. But the CRHR and melatonin circadian rhythms (at least as predicted by our models) became less aligned after physical distancing. This likely represents a less healthy state

Once again, the body is trying to set us up for success. When clocks calculate activity that should occur, circadian clocks anticipate this activity and raise heart rate. At times when activity should not happen, the heart rate slows in anticipation to conserve energy. Still, this needs to be clarified for shift workers. Circadian clocks anticipate activity at the wrong time. This causes extra work for the body, diminishes overall cardiac health, and even predisposes individuals to cardiovascular disease.

Predetecting COVID

When COVID hit, like many other professions, scientific researchers wanted to help fight the pandemic as much as possible. It was novel scientific thinking, such as new mRNA vaccine technologies, that saved millions more lives and eventually alleviated the pandemic. My colleagues wanted to help as well. Even more so, we rely on the public funding of science to make our breakthroughs.

I am where I am today with the Department of Defense, National Science Foundation, and National Institutes of Health because of this. We wanted to pay this forward. So my colleagues and I began brainstorming what we could do to help in the fight against the pandemic.

More than twenty MDs in the Intern Health Study reported that they got COVID early during the pandemic. They reported when they started to show symptoms and what those were. Michael Snyder's group at Stanford also collected a similar dataset and made this publicly available (for which I am thankful). Even more data came from a remarkable collaboration with Choi, whom we discussed in the last chapter. With so many University of Michigan students taking classes remotely, we wondered how the pandemic might disrupt their physiology since many were at home and unable to carry out activities we had previously taken for granted. We asked if any of them would be interested in using a wearable to provide data on how the pandemic might affect them. Amazingly, over two thousand students signed up! More on this dataset later, yet within this cohort, another thirty students were documented getting COVID. With these three datasets, which provided wearable data around the early time of COVID, we compiled a large dataset to study the physiological responses of COVID on heart rate.

Our mathematical analysis of this dataset asked two key questions. The first was whether we could predetect

(detect before individuals show any noticeable symptoms) COVID infection based on wearable heart rate and motion data. The second question we asked was whether it was possible to determine who showed the greatest symptoms. This is especially important for COVID since one of its most unique features is how different individuals respond differently to the virus. Some individuals are asymptomatic. Others get very sick and can show extremely low blood oxygen levels. They might not realize it, but their bodies work hard to stay afloat. Let us tackle the second question here and the first question at the end of this chapter.

The most reported signal that someone might be infected with COVID is increased BHR. This can occur right around the time symptoms start. But what does this mean? One possibility is that they would get stressed as soon as they begin to show symptoms. Stress can increase heart rate, and without proper calibration of stress levels (which is hard to do in the wild), the increase in BHR may indicate the stress an individual is showing. This would greatly reduce the utility of BHR as a "predetector" of COVID. It could just be measuring if an individual is stressed. Would using BHR to detect COVID be more accurate than a questionnaire asking, "Do you think you might have COVID?" Such an experiment has yet to be performed.

We have a better detector of COVID infection than BHR. The body generally does not work harder than it needs to. So any increases in heart rate are typical because of metabolic needs. The heart beats faster to bring more oxygen and nutrients to the body. Yet there is another side to this equation. What happens if the blood has less oxygen to provide? Metabolic demands must be met, so the body must balance this equation by pumping blood more quickly. Thus HR is increased to meet metabolic demands.

We hypothesized that HRpS might indicate who had the greatest symptoms. If someone was symptomatic, we guessed that HRpS might pick this up. Why? If the cardiopulmonary system (particularly the lungs) is not working correctly, less oxygen is being received. When this happens, the heart must beat quicker since metabolic demand must be met. So HRpS might increase during infection since the lungs are not working fully. On the other hand, wearable data collected in the wild is notoriously different to analyze since factors that are not as well controlled in the laboratory could introduce errors. Additionally, it is still being determined exactly how much coughing would reflect changes in cardiopulmonary function.

We separated individuals in our datasets who reported symptoms (in this case, a cough) and those who did not. Indeed, there was significantly increased in HRpS (when scaled for total daily steps as described earlier) for those who reported a cough. This showed that HRpS could measure cardiopulmonary function in the wild within an individual.

We still need to find out all the factors that could affect HRpS. Could other factors also increase HRpS on a daily timescale? Would factors like sodium intake or stress, which increase BHR, also increase HRpS? Much more validation is needed.

Putting It All Together with ML

We previously described some of the effects of COVID on heart rate. To finish this chapter, we consider what heart rate predictors might indicate a COVID infection. The HRpS parameter can increase when the cardiopulmonary function is diminished. The BHR can rise around symptom onset, perhaps due to increased stress. Although it needs further validation, the best mobile predictor of COVID infection might be a diminished amplitude in the CRHR.[8] Circadian rhythms are known to decrease amplitude or even shut off during infection. This makes much sense. When fighting a disease, the body needs to be ready 24/7. Take a break or diminish the activity, and pathogens could multiply, perhaps even exponentially.

So let us return to using heart rate to predetect COVID infection as an example of what our algorithms can do. First, while we describe individual parameters separately, we can use all the metrics presented together to come up with the best estimate of whether an individual has COVID or any other physiological feature that heart rate encodes. Alone, the CRHR seems to predict COVID infection best. Still, with all the parameters, ML can even better predict whether someone has COVID, even before they show symptoms. The parameters we determined played a key role in predetecting COVID. At the time of this writing, though, wearable COVID predetection is far from the level of accuracy and specificity of a polymerase chain reaction test.

Algorithms and ML are continually improving. At the time of this writing, ChatGPT had emerged and was beyond what many of us thought was possible (ChatGPT was not used to create any part of this book). When these words are read, breakthroughs will emerge that are beyond my imagination. In this environment, wearables may likely alert us of infections. May our future algorithmic overlords have mercy on our souls and bless us with knowledge of the future.

Heart, Mind, or Both

Many of us separate the heart from the mind, implying that matters of the heart are less logical, more fickle, and harder to predict. But when we study the heart's rhythms, we discover sublime intelligence. The heart carefully regulates its beat to meet metabolic demands. Not only does it sense this demand; it anticipates the demand, opening shop at times of the day when activity should be high and saving energy when we should be asleep. Heart rate integrates many signals, from stress to light to mealtime and the effects of drugs, which we did not discuss in depth.

Yet how can we understand the language of the heart? We suggested several parameters. BHR can indicate overall health and perhaps even stress. The CRHR tells us when the body thinks activity should occur. The correlation time constant tells us about the effects of hormones like adrenaline. Mathematical analysis allows us to separate our circadian effects from other effects, such as stress. An internal noise amplitude measures these extra effects.

But heart rate is important for activity too. It is a useful metric for athletes wanting to train to peak performance. BHR and HRpS may be lower for these athletes. As wearables are refined, individuals will use new activity metrics beyond steps. The future holds promise with alerts for when we are stressed, fighting an infection, or are desynchronized from the solar day. This will translate into better performance, less cardiovascular disease, and better mental health outcomes.

The real promise of heart rate is the ubiquity of the data. Other vital signs may have less variability, but heart rate is easily measured. Your phone now has thousands of heart rate measurements, and with these large numbers of measurements come more accurate insights into physiology. Hospitals are now collecting wearable data, which can translate heart rate patterns and predetectors of many diseases. Central to this is the ability to separate heart rate signals into multiple physiological signals. These signals can also be combined with those from other physiological time series, such as temperature and sleep, to alert us at home about potential adverse health consequences. When this happens in the future, we will drive straight to the hospital, our phones will unlock the door to the examination room, and the physician will have days or even years of minute-by-minute data.

MOOD

It was late on a Saturday night many years ago. I had gone to bed quite purposefully. On Sunday mornings, I worked at a church, playing their pipe organ and directing their choir. I needed the job, both financially and personally. The phone rang. It was a friend who was upset. At least in our previous conversations, she had a logical way of looking at the world. But this time was different. She knew I needed to sleep, yet it was clear that there was a much more serious problem. Then came a long, animated discussion of a sinister plot involving someone who was famous and a member of well-respected institutions. The more I explained I needed to get to bed, the more she became animated. Knowing I was about to give a public performance in a few hours with too little sleep, I suggested we talk about things tomorrow. By then, however, she had acted on these thoughts in a way that would strain her relationships with others for many years.

My first encounter with a manic individual would not be my last, but it left a lifelong impression. Reading books by Kay Jamison about herself or the poet Robert Lowell describes similar episodes of wonderful, talented people where bipolar gets the better of them.[1] But what could I do about this?

Shortly after arriving at Michigan, after giving one of my first talks on campus, a well-dressed man in a bow tie approached me. Melvin McInnis, a leading figure in the study of bipolar disorder, asked me why I did not work on manic depression. It was a rhythm, after all, and that was my specialty. He was right; there are many fascinating rhythmic features of bipolar disorders and mood in general. But rhythms in mood are much more complex to understand than other ones. Rhythms reflect unpredictable events rather than an underlying timekeeper. Even defining mood can be complex. Here we will focus on two kinds of rhythms in mood. First, we will study rhythms of mood in bipolar disorder. We will then show how all people have rhythms in their mood as determined by internal circadian timekeeping and clocks timing seasons.

How to Measure Mood and Why

Mood is a much richer concept than one might naturally expect, beyond a simple determination of whether we are happy or sad. For example, mood could include the concept of volition or whether individuals feel empowered to take action. A similar but distinct quantity is energy. We typically think of happy individuals as having more energy than those who are sad, but this is not always the case. Concepts such as anger and frustration can also be important factors in mood. Similarly, rumination and irritability are crucial factors. When something bad happens, do we focus on it or let it pass? Substance abuse disorder is often found with bipolar and other affective disorders. Should that be measured as well? Could these ideas affect mood measurements?

The extremes of mood are much easier to recognize. Severe melancholic depression, where an individual cannot complete even the most basic daily tasks, or someone deep in a manic episode, where credit cards are spent out, could trigger interventions or hospitalization. Still, for most of us, mood affects our quality of life and key decisions. Like other body rhythms, some simple mathematical concepts can help us understand mood. But before we dive into those, it is important to understand how mood is measured.

At the highest level, mood can be measured in a simple, one-response survey. For instance, rate your mood on a scale of one to ten. Perhaps better than numbers is to present an individual with several drawn faces ranging from elation to anger to despair and ask them which best

But rhythms in mood are much more complex to understand than other ones. Rhythms reflect unpredictable events rather than an underlying timekeeper. Even defining mood can be complex.

corresponds to their current state. In real-world studies with thousands of individuals, asking anything more might be too burdensome for people to do regularly.

Slightly more complex are questionnaires, including the PHQ-9 for depression or YMARS for mania. The PHQ-9 has nine questions scored on a twenty-seven-point scale, asking you to rate things like "feeling tired or having little energy," "little interest or pleasure in doing things," or "feeling down, depressed, or hopeless," with higher numbers indicating depression. It more precisely measures mood, and starts to get at higher-level concepts like irritability or energy levels. This is not hugely burdensome and often clinically used. If one wants a time course, a subject can take the PHQ-9 or other questionnaires at different times, but taking it multiple times can become burdensome. Even the PHQ-9 might not fully measure mood. Some may also associate mood with boredom, contentment, or fear.

Passive mood sensing would be the least burdensome and provide the most measurements over time. Moreover, it could rely directly on physiological data rather than self-reported answers, which might be more accurate. But what are the physiological signals of an elevated mood? A higher heart rate might mean increased energy levels. As we will discuss in the next chapter, we can directly measure brain activity through EEGs, and these signals might indicate mood. Another promising avenue comes from speech.

Clinicians have mentioned to me that they can tell if a patient is depressed, such as by the cadence, pitch, and so on, of the patient's speech. The words might also be helpful, as would simply asking them if they are depressed, but not accessing words lessens privacy concerns. Emily Mower Provost developed an app that can analyze speech to extract features like cadence, which could not be used to determine what was said but instead could be predictive of mood.

So while determining how to measure mood is important, determining how mood changes over time is a key question too. This will be a crucial focus of our discussion and holds great promise. What if there was a way to tell if someone was about to enter a manic episode? Could that have triggered my friend to pause that night? Could that have helped her seek treatment? Could a time course of mood have helped physicians to determine the optimal treatments?

A Framework for Mood

Transitions between mania, euthymia, and depression in bipolar subjects are frequently thought of as being random. But mood is not just about fluctuations or random events in life; it is about how we react to these events. What is it that makes a mood episode different from a mood fluctuation? Suppose an individual had some life event that decreased or increased mood. Perhaps a rejection or unexpected acceptance. In such cases, an individual is not "depressed" or "manic." Depression and mania have sustained qualities to them that are not simply driven by external cues. When an individual is triggered, their mood will tend to stay in this state for much longer than a normal transition.

Mathematically, the sustained quality of mood episodes can be represented by differential equations—the ways we mathematically represent how systems behave over time. The formalism goes back to Isaac Newton, who developed them to explain the laws of motion. You give a differential equation of the state of a system, and the result tells us the "differential" or how the system's state will change over a minimal amount of time. The system can then be updated in time, and the processes repeated. With a differential equation and some initial state, one can determine how the system will behave at future times if some mathematical conditions are met and a method (typically computational) is used to solve the equations. Historically, these equations were solved by hand, similar to how you would solve an algebra problem. Only a select few are solvable this way, but computers come to the rescue, and most differential equations are easily solved with computers.

Our model, however, also needs to account for the fluctuations of mood resulting from normal life, so a single solution will only work for some scenarios. Luckily, mathematical luminaries like Kiyosi Itô, Henry McKean, Norbert Wiener, and others have worked out the mathematics of "noise," "fluctuations," and "uncertainty," which allows differential equations to cover a range of unforeseen circumstances. This work has many fascinating properties. For example, in the world of noise, as time proceeds forward, it yields fluctuations on the scale of the square root of the time that has passed. We call these equations "stochastic" since they account for these fluctuations. Two additional parameters are then needed for our stochastic model. The first is the noise level or *mood variability*. Moving, changing jobs, and many unpredictable events occur, and this parameter might be high. If life is boring, perhaps these fluctuations are small and the noise level low. The second parameter indicates how quickly the system will respond to the noise over time. We view this as measuring the *neuroticism speed* of an individual. Slow neuroticism speed and a person will have long lasting responses to the unpredictable ("noisy") factors that affect mood.

The behavior of the model was similar to that found in individual bipolar patients. Mood will typically fluctuate around three values, or fixed points in differential equations parlance: one for mania, one for depression, and one for euthymia (normal mood). These values can be personalized to an individual. Yet every so often, the noise in the model will push the system from one region to another, such as from mania to euthymia or from euthymia to depression. You can start the model in any state: mania, euthymia, or depression. Fluctuations, though, can push and eventually cross the thresholds between E to M or E to D, and a mood episode begins. A bright undergraduate student, André Schultz, who also was an Olympic-level swimmer and trained with Michael Phelps, put this all together into a mathematical model.[2]

Testing and Refining the Framework

A natural question, then, is whether there is a clock driving mood episodes. After Amy Cochran, now a professor at the University of Wisconsin, looked carefully at McInnis's data, there was no clock (at least that we could detect mathematically) in bipolar driving transitions between mania and depression. Its rhythms fluctuate like the flipping of a coin. This was useful since it matched our stochastic differential equation framework. She could not rule out the possible rare subject who was rhythmic; the dataset certainly did not contain all bipolar subjects, so we could only speak generally.[3] Also, as we will see below, on the timescale of a day, mood is regulated by the circadian clock.

Amy had access to data from surveys measuring mania and depression, such as the abovementioned PHQ-9 (depression) and Altman (mania), which the subjects filled out from time to time. We expected an almost perfect downward correlation: when mania was up, depression would be down, and vice versa. This was not true, nor were they independent (see figure 6). When mania was up, depression tended to be down in most subjects, but in some, they were up. Psychiatrists call these states *mixed episodes* in which people might, for example, have the increased energy of someone in mania and the sadness typical of depression. As we touched on at the beginning of this chapter, mood is a complicated thing to define. After some tweaking, Cochran settled on a somewhat modified version of the equations we used. It turns out they had been used before by financial mathematicians studying interest rates. Thus we used the Cox–Ingersoll–Ross model from mathematical finance and modified it to bipolar data.[4] Who would have thought that interest rates and mood in bipolar subjects are regulated in similar manners?

Putting it all together, if we had a history of mood on a patient, Cochran could fit the parameters of a model to their data. The parameters of this model could then be used to help determine if the individual was normal or bipolar, and then more specifically, if the individual had bipolar I, bipolar II, or other classical clinical classifications. The DSM, the official diagnostic manual clinicians use to diagnose mental illness, sets the gold standard clinical guidelines. But then we began to wonder if the gold

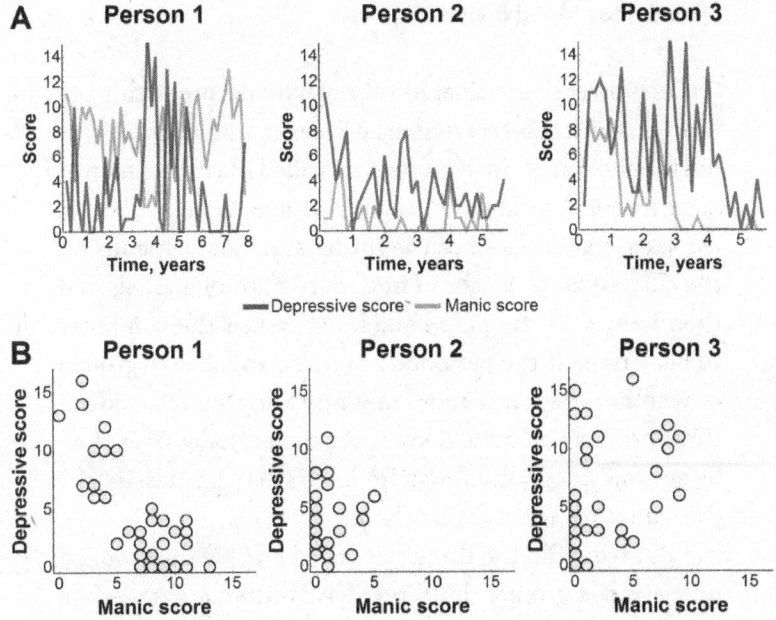

Figure 6 A. Time course of mood for three bipolar individuals. B. Many times, individuals show features of both mania and depression. *Source:* Amy L. Cochran, André Schultz, Melvin G. McInnis, and Daniel B. Forger, "Testing Frameworks for Personalizing Bipolar Disorder," *Translational Psychiatry* 8 (2018): 36.

standard clinical classification was optimal. Perhaps in the future, models will be able to classify bipolar disorder better. The DSM's parameters are also numbers without hard thresholds. So they would naturally describe the spectrum of behaviors seen.

A Model for Mood Switching

So Cochran then considered a simpler model predicting the rates at which subjects switch between manic, euthymic, or depressive states. Such models are called Markov—named after Andrey Markov, a Russian mathematician who studied such systems—since the future state just depends on the current state. Cochran fit data to Markov models and then looked at the parameters. She asked the computer to determine if the parameters tended to fall into groups or whether they were more randomly distributed. Indeed, three groups emerged. Looking at the episodes over time by eye was not good enough for us to see these classes, but the computer could detect them.

But what did the three classes mean? We then looked to see if the groups could predict additional data—data that was not used to form the group. One of the classes had double or more suicide attempts than the other ones.[5] In this manner, classification gave a way to look at the rhythms of mood episodes and determine who was most at risk for suicide. Much more testing needs to be done before this can be fully used in the clinic. Still, all signs point to a future where these algorithms give a more precise personalized treatment of bipolar disorder, particularly regarding who is at most risk for suicide.

We hope that clinicians can integrate these mathematical tools into clinical practice so that mood time courses can be tracked and analyzed using our methods, and that a psychiatrist can use these to know who is most at risk. But perhaps most important, this work shows how rhythmic disorders like major depression and bipolar are part of a spectrum. If we can estimate parameters, we can see where an individual is on the spectrum. They may not officially meet diagnosis criteria but are close to the border. They may be subthreshold and yet could still benefit from some treatment. Moreover, parameters might change over time. Hence strict diagnoses could be replaced by risk factors or probabilities, which could more precisely predict the future course of an individual's mood.

Circadian Rhythm in Mood

Understanding and predicting suicide, or episodes of mania or depression, is crucial. But what about the rest of us who do not have affective disorders? Could we detect patterns in mood that are not pathological? Could understanding fluctuations in mood also help us cope with the general stresses of modern life? Like many other signals we have studied, a good place to start looking for rhythms in mood is to look for circadian rhythms in mood. There are two ways to do this—one focused in the lab and one in the real world.

The circadian clock controls the general excitability of the brain as well as overall energy levels, sleepiness, and metabolism, all of which can contribute to mood. Yet this is different from what individuals think when they think about mood rhythms. We rarely think about our circadian rhythms, just as we rarely think about our heart rate, because they are both unconscious endogenous parts of our body. We do not consciously know the state of our circadian clocks even though we can feel the effects of this timekeeping. We know how long we have slept, at least to a more accurate degree than circadian timekeeping. Everyone knows that sleep affects mood. Many of us have known someone tired and irritable (perhaps our spouses have seen it in us). Sleep timing and fatigue play an important part in mood too. But do circadian rhythms affect mood beyond this?

Diane Boivin examined this twenty-five years ago in a landmark study published in the *Archives of General Psychiatry*.[6] She recruited subjects to live on a non-twenty-four-hour day in a similar experimental setup to the experiments discussed earlier. Living on a non-twenty-four-hour day dissociated their circadian rhythms from their sleep-wake cycle. When she found circadian mood rhythms in individuals who were not sleep deprived, they were not that large. But because of the longer than twenty-four-hour days she used, she was able to measure mood in individuals with sleep deprivation, as individuals stayed awake for more than sixteen hours. As individuals became more sleep deprived, the circadian rhythms in mood became much more prominent. Sleep deprivation increases the amplitude of the circadian rhythms in mood. If the circadian clock suggests a lower mood and an individual is sleep deprived, their mood becomes much more depressed.

While this was a heroic study, experiments where individuals live for many days in time isolation at a clinic or hospital are costly; only a limited number of subjects could be studied. In some sense, this kind of laboratory study is better than the real world since many potential confounders of mood are removed (it is a world free of rejection, unexpected joys, etc.). It is a significant first step. One wonders, though, if these key effects will still be seen in the real world. This led to recent work with the Intern Health Study.

Interns in the Intern Health Study work on night shifts that induce sleep deprivation and shift circadian rhythms. They get queried once a day, typically around 3 p.m., to fill out the one-question mood survey mentioned earlier. Yet some fill it out later or earlier in the day because they might be too busy working with patients or even asleep after a long shift when queried. Furthermore, during night shifts, they sometimes fill out the mood survey at circadian phases that are misaligned and with varying levels of sleep deprivation. In total, the study collects over a hundred thousand daily mood assessments per year. Benjamin Shapiro, a psychiatrist at Dartmouth Hitchcock, calculated the CRHR for each day to go along with each of these measurements, and then compiled them to show the relationship between circadian phase and mood assessment among individuals and across the population. As Boivin found, usually there was a small circadian rhythm in mood. When individuals were sleep deprived, however, the circadian rhythm became pronounced, with low mood scores when both sleep deprivation and circadian misalignment were present.[7]

This has practical implications. When looking for predictions of mood levels within a day, individuals should be particularly wary of the combination of sleep deprivation and circadian misalignment. I sensed this personally on a recent trip to Heidelberg, Germany, where I gave a keynote talk at the Society for Mathematical Biology, which met with the European Society for Theoretical and Mathematical Biology. This opportunity to give such a keynote only comes rarely in a career, so I wanted to give the best talk possible. I flew to Europe, but the night before the talk, I was convinced it was no good. I would be in front of the entire field and worried that the evidence I had was insufficient. Perhaps I was a failure? Alone in a dark hotel room, I called my wife. She carefully listened to all the exaggerated worries and depressive talk. Finally, she asked me how much sleep I had gotten. I replied, "With so much work, I only slept for about three hours for the past couple nights." Then she asked me to look at my group's app to see what time my body thought it was. Sure enough,

I had entered this region of sleep deprivation with circadian misalignment that can depress mood. I hung up, got some sleep rather than worrying, and gave what I thought was one of my most successful talks the next day. I would like to say my research saved my talk, but perhaps it was my wife.

Mood and Circadian Desynchrony

While sleep and circadian rhythms affect mood, the Intern Health Study offers many additional opportunities to understand mood factors. Yu Fang, one of the permanent researchers at the Intern Health Study, sought to find these factors. As we described earlier, the effects of sleep deprivation on mood are complex and have a circadian dependence. But beyond circadian rhythms, what other factors can predict daily mood scores? Fang discovered that sleep deprivation was just part of the story.[8] As we noted above, typically sleep deprivation does not always have a large effect on mood, especially when it comes at circadian phases that hide its effects, such as when the body thinks it is day. The top predictor of mood in her analysis was something called sleep variability. Seen from the opposite view, consistent sleep timing each night greatly enhances mood. This joins a recent finding as this book goes to press that circadian phase shifts, as predicted by the mathematical models discussed in chapter 3, are a leading predictor of mania or depression, and that when sleep occurs at the wrong circadian phases, large changes in the PHQ-9 can result.[9]

There are many potential reasons why someone would have variable sleep timings. Different schedules and circadian phases for the interns would affect when they go to bed as well as how long they sleep. If they try to sleep at the wrong time over several days, effects could compound, sleep will become more variable, and you will reduce sleep quality. Even if sleep occurs, it could be less restorative if there is circadian misalignment. Additionally, individuals with more stress may be unable to sleep and have more variable sleep. There might be a feedback loop. Stress might cause less sleep, which exacerbates the effects of stress. Variable sleep might also indicate poorer general health.

This work joins a larger body of evidence indicating that regular sleep timing is essential for general health. For example, individuals in Germany tend to sleep in on weekends, shifting the sleep cycle by several hours each weekend. Such changes in sleep timing could trick different body parts into thinking that it is different times of day. Our digestive system could align with mealtimes, whereas melatonin could try to predict when light should occur and our hearts when we exercise. As the body tries to coordinate activities, it could get confused. This could be worse than if all the clocks were aligned but at the wrong time.

The Circadian Hypothesis of Bipolar

But more than the circadian rhythm of mood, some believe that circadian rhythms might be the cause of bipolar disorder. Over time, I have become more convinced of this idea for three reasons. The first is simple, though far from any sort of proof. Circadian rhythms have a strong genetic component. Bipolar is highly hereditary as well. Sometimes, studies have implicated the same genes in circadian regulation and bipolar disorder, yet just because these are correlated does not mean they are causative.

Second, lithium is the most prescribed drug for bipolar. What does lithium do? It shifts circadian rhythms in the body (among other potential effects). Lithium targets a key protein, GSK-3β, whose activity shows a strong circadian rhythm. GSK-3β is a signaling molecule within the cell as well as a key link between the molecular clock within the cell and electrical activity of neurons. It regulates how sodium molecules enter and leave the cell, which controls the signaling between neurons. Turn on this current, and neurons signal more. Thus the most important bipolar drug also targets the main link between molecular timekeeping and neuronal activity. Genetics supports this strong link too. Recent evidence from the Broad Institute suggests that a gene related to GSK-3β might be at the heart of the hereditary nature of bipolar disorder.

Third, social rhythms therapy, another key approach to treating bipolar, stabilizes the daily schedule and forces an individual to live a regular twenty-four-hour day. This also aligns sleep with where it normally should be. Social rhythms therapy has been shown to decrease the occurrence of both depressive and manic episodes, and may help to bring an individual out of an episode. Again, affecting circadian rhythms can help treat bipolar disorder. Treating sleep variability and circadian desynchrony helps end mood episodes.

So some of the most effective treatments for bipolar disorder fix or change circadian timekeeping. A good start might be to study how the circadian clock regulates the excitatory nature of neurons. If this got locked into a low state, many of the features of depression might be seen. If one locked it into an excitable state, many of the features of mania might occur. Circadian clocks in the brain might move neurons into or out of these states, and offer a conduit for treating mood episodes. Another mechanism might be through the signaling molecule dopamine. Dopamine concentration is under the control of intracellular circadian clocks. Given this, dopamine might be responsible for some of the phenotypes of mood episodes and a clock signal. These are still hypotheses, however, and need much more work to unpack.

Seasonality and Seasonal Affective Disorder

As discussed earlier, our biological clocks have evolved over millions of years in response to the daily light-dark cycles. Only recently have we moved to a modern lifestyle so separated from the natural environment. If we look at our closest animal relatives, we see that circadian clocks are used to tell the seasons. Rather than using temperature, which can vary significantly between locations, mammals use light-dark cycles to track the seasons. Light-dark cycles can determine when to reproduce, harvest, and hibernate. We typically do not think of human bodies as tracking the seasons, except in the case of mood; seasonal affective disorder, where the shorter days of fall and winter cause depressive symptoms affects millions. There are also reports of seasonality in human birth, death, and infection rates. Typical treatments involve getting brighter artificial light during winter to simulate summer.

But how do circadian clocks track the seasons, particularly light-dark cycles? The SCN has built-in mechanisms to do this through its thousands of cells, each capable of timekeeping. As mentioned in chapter 2, Serge Daan proposed that some cells within the SCN track are in the evening and others in the morning, thereby extending the predicted length of the light-dark cycle. Work by Johanna Meijer at the University of Leiden, though, provides more biological mechanisms for this theory.[10]

Meijer has recorded the activity of neurons in SCN during both winter and summer light-dark cycles. At any time during the light portion of the light-dark cycle, she found some neurons in the SCN that had their peak activity. Some neurons peaked early in the day and others later in the day. As the light portion of the light-dark cycle expanded, so did the time between the phases of the clocks in these cells. If the length of the light portion of the day increased, such as when winter days changed to summer days, the length of the times when we could find neurons in the SCN at their peak activity increased as well.

How is this done within the SCN? Experiments by Jihwan Myung in Toru Takumi's group in Japan, along with modeling done at Michigan by my graduate student Daniel DeWoskin, helped answer this question.[11] Myung hypothesized that clocks within neurons of the SCN are both attracted to *and* repulsed by each other. So imagine two objects and let their physical distance represent their phase relationship. Now put both a spring between the objects and wrap a rubber band around both. It seems strange to do both, but bear with me. The rubber band brings the objects closer together, and the spring does the opposite, bringing them farther apart. More important, the spring and rubber band present a mechanism to set their distance, which is a proxy for how apart their phases are. Increase the stiffness of the spring and the objects move farther apart. Increase the strength of the rubber band and they move closer together. In mice, we found that gamma-aminobutyric acid (GABA) acts as the spring pushing cells apart in the SCN and the length of the light-dark cycle modulates its signaling. Vasoactive intestinal polypeptide functions as the rubber band. Make the spring (GABA signaling) stronger, the rubber band expands, and you have a larger phase relationship and there are active neurons over a longer portion of the day, just as you see a longer portion of light in summer days. Make the spring more pliant and the oscillators come closer together.

So what happens in seasonal affective disorder? Our hypothesis is the SCN does not get sufficient or correct signals about the length of the day. The SCN then does not predict the correct phase relationship between the neurons in the SCN. The body then predicts the wrong season and sleep is disrupted. For some, it might cause us to become more active in winter. For others, it may overcompensate and cause too much of a break in activity in winter. It can explain why, during winter months, individuals might develop depression. Melatonin likely is a key signal in all of this. Too much or too little melatonin in winter, and the body might send the wrong seasonal signals.

Too much or too little melatonin in winter, and the body might send the wrong seasonal signals.

The Many Timescales and Nuances of Mood

While mood can change on the timescale of a year in seasonal affective disorder, there are other long timescales. For instance, consider a phenomena called kindling in bipolar. The idea goes back to Emil Kraepelin, sometimes credited as the founder of the study of bipolar disorder, but who also held some unfortunate views on other topics. The idea is similar to how kindling helps create a fire. First, a small fire on a match gets started. This spreads to kindling, which enables even greater fires in larger pieces of wood. Kraepelin suggested that episodes might initially be rarer and only be seen in adulthood. Yet as time passes, episodes become more frequent, as if each episode primes the next one. In this way, parameters of our models could change with age, such as moving the threshold needed to trigger a manic or depressive episode closer to the average euthymic state.

Treatments for mood disorders take a long time too. One might think of having multiple models for an individual: one without medication, one with lithium, and so on. This tacitly assumes that drugs act fast, though, such as on a timescale of minutes to hours, not the slow transition as a drug takes effect. Selective serotonin reuptake inhibitors and other antidepressants can take months to have effects. Likewise, lithium can take weeks before it starts to show an effect. The mechanisms that create these long-term courses have yet to be determined.

This all means that both the time course of mood and treatment of mood disorders change over slow timescales as well as faster ones. To fully understand the long timescales might take years or even decades of wearable data. Excitability/irritability is accompanied by adrenaline/cortisol, which may pass on the timescale of an hour. Mood naturally varies on the timescale of a day. Mood episodes are closer to the timescale of weeks. Antidepressants may take from weeks up to months to work. Seasonality in mood is a yearly phenomenon. The dynamics of bipolar might change throughout a lifetime, such as predicted by the kindling hypothesis. Once again, a key part of our goal is to parse the different aspects of mood, using different timescales and physiologies as a guide. Mathematical models of mood may resemble the many gears of a grandfather clock, but a complex one that times the hours and seasons, with each gear representing one of these processes.

A Model for Happiness Is Complex

Mood has many interesting dynamics that have yet to be explored. Mood is carefully related to many rhythms. Menstrual cycles affect mood. Sleep affects mood. Circadian rhythms affect mood. Light and light-dark cycles are key, particularly for seasonal affective disorder. Mood is built on rhythms that are built on rhythms.

This all means that both
the time course of mood
and treatment of mood
disorders change over
slow timescales as well
as faster ones.

Moreover, the data on mood takes many different forms. Genetic data from individuals with mood disorders can yield insights into behavior. We can track mood in many ways. Along with the tracking, wearable data provides physiological measurements of what could happen inside an individual. These could extend to a wearable EEG, which could measure brain processes directly, as described in the next chapter.

Some of these rhythms time events, but others, like episodes of mania and depression, are inherently random. This is a crucial point to consider. We should be able to predict when a high or low mood, or a mood episode, is *more likely* to occur. We can diagnose mood disorders in the future through fitting parameters. While models will play an increasing role in understanding mood dynamics, they will never be predictive enough to exactly time when the next mood episode will appear. But they can quantify risks and warn of the possibility of mood episodes. Such predictions help guide us to therapies or even simple suggestions, such as when to take a nap or go outside for a walk in the park.

Like my late-night experience in a German hotel room, we might be able to attribute much of our mood to physiology. I felt helpless and depressed partially because of my sleep drive and circadian state. Furthermore, affective disorders can be understood through physiology, proteins

While models will play an increasing role in understanding mood dynamics, they will never be predictive enough to exactly time when the next mood episode will appear.

like GSK-3β, or mistimed signals like melatonin. For me, this makes a critical societal point. With the cause of these disorders being physiological and explainable, and even in some cases predictable with mathematical models, society can approach these disorders differently, removing some of the harmful stigmas surrounding these mental health challenges.

BRAIN RHYTHMS

Soon, measuring electrical signals from the brain will become commonplace. EEG measurements from wearables could be as prevalent as heart rate or activity measurements from other wearables. Measuring signals from the brain could be the most informative signal we study. Yet brain rhythms are the most complex; models and analysis are essential to decipher these signals. They also can present the greatest ethical challenges.

Recording the Electrical Activity of the Brain

There are many ways to measure the activity of the brain. The most common is an EEG. Subjects typically have professionals place electrodes on their scalp. These electrodes can pick up signals from the cortex, the part of the brain closest to the scalp. Different electrodes record different brain parts based on where they are placed. Each EEG electrode records from tens of millions of neurons and cannot resolve the behavior of individual neurons but instead give an overall picture of a brain region. The part of the brain closest to your forehead is the prefrontal cortex. It is a region where many of a person's thoughts are processed. In the back of the head is the visual cortex. It is a region where signals from both retinas in the eyes end up after initial processing. Both will signal differently depending on whether you are deep in thought or your eyes are open. Auditory sounds are processed in the auditory cortex, about halfway between the visual and prefrontal cortex.

Typically, sixty-four EEG electrodes are used in research settings. The electrodes are sometimes put in a gel when attached to the scalp to get the best signal, allowing the electrical signal to travel better. Recent work, though, has developed dry EEG signals that are working with increasing accuracy. Dry electrodes have now been incorporated into headsets that can still measure brain activity, making EEG recording as simple as wearing a hat. My experience with them suggests they can be relatively comfortable—much more so than having electrodes permanently implanted in the brain.

Wearable EEG headsets typically have fewer electrodes than a clinical EEG; eight is a lot for a wearable headset. The recordings are noisier. These wearable headsets are

Brain rhythms are the most complex; models and analysis are essential to decipher these signals. They also can present the greatest ethical challenges.

starting to be widely used, such as to help with meditation, aid coders to concentrate better, score sleep, and even detect fatigue. In the future, physicians could also use them to diagnose diseases, particularly in underresourced communities. If the students in a classroom all wore EEGs and shared this data with the professor (which would never happen), the professor could see who was paying attention and who was dozing off.

Beyond the EEG

While we will focus this chapter on the EEG, other techniques for monitoring brain activity exist. Functional magnetic resonance imaging (fMRI) is a more intensive way to measure the brain's activity. It detects blood flow, and indirectly, the electrical activity of neurons. As neurons signal more, they need more blood flow; a scanner can detect this. The difficulty with scanners is their cost (a million dollars or more) and the fact that an individual must be in a fixed supine position. The number of activities that subjects can do in a scanner is limited. Nevertheless, it gives a much more precise measurement of the brain's activity. Scanners can record activity from a hundred thousand or more brain regions, over a thousandfold more than an EEG. These regions can be deep within the brain, such as the SCN, unlike an EEG, which only measures electrical signals close to the scalp. Yet the temporal resolution of an EEG is typically about a thousandfold more than an fMRI. That said, an fMRI can measure many different things too. New scanners can visualize the fibers consisting of neurons. They can even measure signaling molecules like GABA. In one fascinating experiment, subjects were played movies, and their fMRI was recorded. Researchers then decoded which part of the movie they were watching from the fMRI signals.

Another method that has the benefits of both an EEG and fMRI is near-infrared spectroscopy (NIRS). NIRS measures oxygenation in the blood, which, similar to what is used in an fMRI, is a proxy for neuronal activity. This works by shining light into the brain and recording light that is reflected backward. The benefit of NIRS is that it can be miniaturized to be placed on an individual without the constraints of a scanner. It is also much less expensive than a scanner, although more expensive than an EEG. NIRS contains several techniques, one of which is to study the time domain signals.

As I write this, brain-computer interface is one of the fastest-growing areas of computer science, garnering the attention and investment of Elon Musk and others. We can now directly record from neurons in the human brain with high accuracy with implanted electrodes. One example of when this is useful is deep brain stimulation. For severe cases of Parkinson's disease, electrodes are implanted in the patient's brain and provide a signal that can reduce tremors. Deep brain stimulation has also been used to treat epilepsy, depression, and even obesity. On the way to implanting the electrodes, physicians listen to the neurons to determine the optimal placement so that the stimulus will have the maximal effect. These live recordings give a near-exact picture of neuronal activity in the human brain. The electrode can rest next to a neuron and pick up individual spikes attributed to individual neurons. But for most of us who do not want Musk's devices in our brains, we must settle for other measures.

Fourier Series and All That

From now on, our analysis will focus on EEG signals for simplicity. There are many timescales to these signals. Changes in electrical activity can be measured on the timeframe of fractions of seconds to sleep patterns on the timescale of a night. Similar to what has been described in other chapters, these different timescales represent different physiological signals. What we need is a method to extract signals with specific frequencies. The way to do this goes back two hundred years to French mathematician Joseph Fourier.

Consider any measured periodic signal—that is, a signal that returns to the same state that began within a set time we call its period. EEG signals have this property. While there are an infinite number of possible signals like this, one periodic signal has received the most study. It forms the basis of much of our knowledge about periodic signals: the sine function. The sine function has a counterpart, the cosine function, which is the same as the sine function, just shifted in time one-quarter of a cycle. Together they have many unique mathematical properties. As time progresses, if you plot points where the sine function's value is on the horizontal axis and the cosine function's value is on the vertical axis of a graph, you will get a perfect circle. The amplitude of this circle never changes, and thus no specific time holds a special place, just as no time of day holds a special place on a wall clock, and time progresses at the same rate from one minute to the next. This is one of the features that make sine functions so useful. Another is the following. Shift the sine function in time by any amount. You can now represent the new signal as the weighted sum of the original sine function and its cosine function as long as these original functions are appropriately scaled. But can we go further? Now consider any periodic function. Fourier demonstrated a unique feature of any periodic signal: it equals the sum of a combination of sine and cosine functions. Hence any periodic signal can be represented by sine and cosine functions, which are unique in that they do not prefer any particular phase over another. Fourier also showed how to take a periodic signal and find how much of it can be explained by sine and cosine functions of a particular frequency (the Fourier transform).

As such, we can take our EEG signal (or any other periodic signal) and represent it as a "Fourier series," a sum of sines and cosines. We now have efficient algorithms (fast Fourier transforms) to extract a sine and cosine function that best represents the behavior of any periodic function at a given period. So we take our EEG signal and remove the part of it (the component) with a period of one or half a second, and so on. This forms the basis of how we analyze the EEG.

Bands, Waves, and Neuronal Responses

Talk to an EEG expert, and you will quickly hear Greek letters as if the expert is in a fraternity or sorority. This is not a nod back to their college days. Instead, the standard way to analyze the signal from an EEG electrode is to look at its component with different periods (or in their terminology, frequencies, which are just one divided by the period). Frequency ranges are named with a Greek letter. Sometimes the EEG has many frequencies superimposed on each other and looks chaotic. Sometimes the EEG stops its chaotic behavior, looks reasonably periodic, and shows one preferred period. This depends on the state of the brain. We will briefly describe the signals in specific frequency bands called waves, where the frequency is assigned a Greek letter (see figure 7).

Gamma waves are the fastest wave typically studied in the EEG. They can have a frequency of 100 hertz (Hz) or even a bit faster. When gamma waves are present, individuals are in states of deep concentration. Gamma waves indicate attention, such as to some visual stimulus.

Beta waves, EEG activity from about 15 to 30 Hz, are fast signals typically seen when an individual is awake. They change rapidly and look stochastic. Some neurons can send signals at this rate by themselves. These neurons are typically found deep in the brain, however, and are not part of the cortex, the part of the brain closest to the scalp. Beta waves represent the combined and perhaps disorganized activity of many neurons. These are typical for awake individuals.

Alpha waves are the EEG signals with a frequency of around 10 Hz. This is an important number since cortical neurons typically send signals around this frequency. Alpha waves can be large—the largest commonly seen in an EEG. They are best seen when an individual shuts their eyes and thus are the base state when visual information is absent. Alpha waves are also seen during REM sleep, and meditation can increase their amplitude.

Theta waves are EEG signals with a frequency of about 4 to 7 Hz. When strong theta waves are seen, an individual

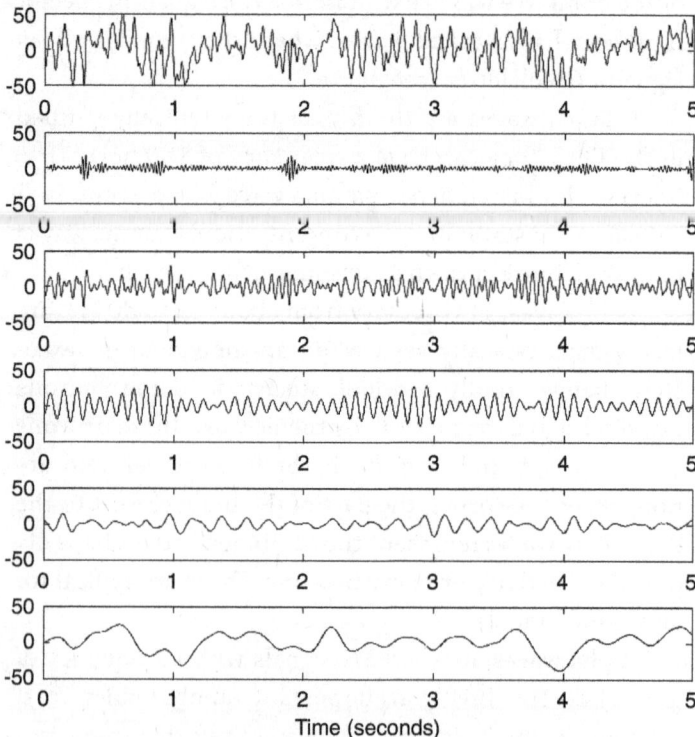

Figure 7 An EEG recording from the author. The original rhythm is plotted on the top, followed by the gamma, beta, alpha, theta, and delta signals.

is likely in lighter sleep or deep meditation stages. These waves are also associated with the hippocampus, a brain region important for memory formation.

Delta signals are the slowest waves typically considered. They have frequencies around 1 Hz. Delta waves are only seen during the deepest stages of sleep, NREM. As we discussed in the slow wave sleep section of chapter 3, neurons in NREM sleep briefly enter silent periods when their signaling shuts down.

In addition to specific frequency bands, there are some signature signals that researchers typically look for in EEG signals. Rather than looking for intrinsic behaviors at a particular frequency range, a specific environmental signal is given to an individual and the response in the EEG is recorded. These, too, can indicate overall brain health and act as clinical markers of disease. One example is the p50 protocol. This consists of playing a sound and recording the response on the EEG 50 milliseconds later—the typical time it takes the brain to register that sound. Scientists can repeat the sounds half a second later to see how the brain has adapted to learning those sounds. Another protocol uses speech and gives an individual a word or phrase that does not make sense. This takes longer for the brain to process, so in the p600 protocol, the researchers wait for 600 milliseconds and then record the response. These are only some kinds of stimuli that can be used. Individuals can even be presented with movies and the responses recorded.

Specific physiological signals underlie the behavior at different frequencies. These behaviors can be due to single-cell, network, or even whole-brain-level activity. Computational and mathematical modeling is needed to understand exactly where these behaviors originate. Only recently have we had the tools to simulate the brain at scale and represent individual neurons' complete physiology. We are just beginning to develop the tools to explore these behaviors.

Rhythms of a Single Neuron

Individual neurons could generate some of the electrical behaviors seen in the EEG. Luckily, the electrical activity of a neuron is well understood through the work of Alan Hodgkin and Andrew Huxley going back about seventy-five years.[1] Hodgkin and Huxley studied the behavior of the squid giant axon, with the axon being the most electrically active part of the cell. Squids are an ideal animal for studying neurons since they have two of the largest "giant" neurons found in nature. They are so big that you can see them with the naked eye, and this allowed Hodgkin and Huxley to experiment with them. Hodgkin and Huxley proposed that the electrical activity of neurons was generated by a cell carefully letting ions like sodium and potassium pass through the cell membrane. This generates currents that can signal to other neurons. This basic physiology holds whether you are considering a neuron in the squid, mouse, or human. What causes the difference between the activity of individual neurons or neurons in different species is which currents are used. So while a squid might use chloride, at least as proposed by Hodgkin and Huxley, a mammalian neuron might use calcium in its place. Additionally, the way calcium flows into the cell might be different between neurons in an animal and comparable neurons across species.

Individual neurons can all individually generate behaviors that can be seen on an EEG. A few neurons might generate the rhythms, but these are then transmitted to the rest of the brain, which responds accordingly. Change the currents within a neuron (possibly by turning on genes) and the frequency of the response would change. This is different from network-level behaviors, where neurons collectively generate rhythms like the "wave" at a football or baseball game. Network-level behaviors are described in the next section.

An individual neuron is capable of a wide range of behaviors. The neuron can be silent. The neuron can signal. When it does, the standard way it signals is through an action potential. Using perhaps overly dramatic language, neuroscientists typically say that a neuron "fires" an action potential when the signal is sent. During an action potential, the neuron briefly raises its voltage, only for a couple milliseconds, but this is enough of an electrical signal for it to be felt at the ends of the neuron, where two neurons almost touch. These parts of a neuron are called synapses. While the voltage is raised, calcium, a signaling molecule in the cell, flows into the neuron, which causes other signaling molecules to leave the neuron. Those signaling molecules can either inhibit or encourage (excite) the signaling of the other neuron at the synapse, depending on which molecule is released. GABA is the critical inhibitory molecule. Glutamate is a possible excitatory signaling molecule. But lots of molecules can signal neurons. They can even have special sensors for melatonin, for example.

But more complex patterns exist. Neurons can enter states where they repetitively signal (fire) action potentials to other neurons. Individual cortical neurons can do this at around ten times a second, putting them in the alpha wave band. Neurons can also enter a bistable regime where they can either fire action potentials repetitively or stay silent, depending on what signals they receive. They do this like a light switch. Give them a brief signal to turn on signaling and they can fire action potentials as long as they like. Get tired of signaling and send them an appropriate signal, and they will stop all action potentials. What are these signals, though?

This starting and stopping of rhythms of action potentials is important for many processes. David Paydarfar is particularly interested in these signals.[2] A neurologist by training, Paydarfar recruited me as a graduate student to study this signaling at the Marine Biological Labs in Woods Hole, Massachusetts. These labs were the last places in the United States (that we knew of) to collect squid for scientific research on electrical activity. John Clay, a researcher at the National Institutes of Health who spent his summers at the Marine Biological Labs, worked with us too. Now retired, Clay was the leading squid giant axon experimentalist at the time. We found that neurons look for rhythmic inputs to determine whether they should start or stop their output rhythms of action potentials. Yet the effective inputs could vary. Some looked at growing oscillations; others had oscillations with a significant negative current. What this means is that even on the single neuron level, rhythms in electrical signaling follow a complex language we do not fully understand.

Neurons can also signal in ways beyond their action potential. They can signal through their resting membrane potential. This is like the resting heart rate we studied earlier. After an action potential, the neuron returns to a resting voltage. For most neurons, this is around minus-sixty millivolts. But sometimes a neuron rests in an excited state. SCN neurons do this, as we describe next. They have an excited resting membrane potential and can signal to other neurons without action potentials.

We presented this at the biannual Society for Research on Biological Rhythms meeting (the Oscars for clock research), expecting boos because it was controversial and had not been seen before. We did find the boos, and many felt this was not possible. Late one night, however, a graduate student of mine, Casey Diekman, banged on my door. He told me that earlier, at a party, he had met another graduate student, Mino Belle, who had discovered these states experimentally in the SCN. Belle, too, encountered resistance and needed a model to help put it all together. As soon as we returned home, I suggested Diekman fly to Manchester for further exploration. Yet with grants and funding always scarce, we could only find a flight from Toronto that was cheap enough. Diekman borrowed my car, drove to Toronto, and flew to Manchester. Before we knew it, we had a paper published in the journal *Science*.[3] Now Diekman and Belle are both professors.

These are some of the potential behaviors of a single neuron. Neurons can also show bursting firing when several action potentials are separated by a period of silence. As well, neurons can show chaotic firing where action potential appears at seemingly random times. Indeed, the behaviors of single neurons are vast. But this is only the smallest part of signaling in the brain. Much more can be seen through the collective behavior of neurons.

Short- and Long-Range Connections

While the brain may seem like an amorphous collection of neurons, it has a clearly defined structure. Neurons close to each other form local, sometimes dense connections. This is especially true in the cortex, the top part of the brain, which writer Mark Twain thought looked like a cauliflower. The cortex has a clearly defined structure of layers, and in general, there is a typical pattern in how these layers are connected. Local groups of neurons form units that can perform specific computations. For example, the visual cortex is organized into columns that process visual information from a part of the visual field. Connections within these local regions are called gray matter.

Then there are also long-range connections between different brain regions, which are called the white matter of the brain. Connections in the brain can be rather long. Individual neuronal projections can be as long as inches, even though the neurons are too thin to see. For example, starting at the retina, in the front of the head, signals travel down axons making up the retinohypothalamic tract to the hypothalamus, which we discussed before. Eventually, signals are processed in the visual cortex at the back of the head.

This combination of gray and white matter has caused people to consider the brain to be connected as a "small-world network"—an idea suggested by Steven Strogatz and colleagues. A small-world network can be generated from dense local connections by changing some of these local connections to long-range ones. Small-world networks are good at transferring information between different neurons. If we instead chose random connections, the local regions could not perform their local functions, like analyzing a particular part of the visual field. But if it were all local connections, information could only travel from neighbor to neighbor and thus slowly throughout the brain. Much of the information encoded in an EEG consists of signals traveling between different brain regions, such as to try to synchronize firing across regions, requiring long-range connections.

Connections could be inhibitory, meaning that signals can prevent a neuron from signaling, or excitatory, where signals bring a neuron closer to firing. Having inhibitory connections greatly increases the range of possible behaviors too. For instance, consider a group of neurons connected only by inhibitory connections. For low signaling strengths between neurons, they would just signal close to what you would find if they were not connected, with disorganized firing. As the connection strength increases, however, the behavior differs from what you would expect. Neurons "cluster" and form groups of neurons that fire together. These clusters—be they three, four, or more—all signal out of phase. So a group of neurons signals, and then another, each taking turns.

This is one of many possible behaviors seen when neurons are connected. Another behavior often attributed to cortical neurons is interneuron gamma. Here, inhibitory neurons signal when no excitatory input is present. Their firing stops other cells from firing until the next round of bursting. They are almost like the gatekeepers at a children's amusement ride, allowing some children to enter (signal), and then more important, making others wait until the next time the ride runs. A variation has been attributed to cells in one specific cortex layer, the pyramidal cell. Hence pyramidal interneuron gamma, which differs slightly from interneuron gamma since excitation is required to get neurons to get the inhibitory neurons to signal.

But this is the beginning. Pyramidal interneuron gamma, interneuron gamma, and clustering can all be seen with just a few neurons, and are easily understood. Reading the literature, one is tempted to think that these explain all phenomena, which appeals to some. Still, many possible behaviors remain to be discovered when considering the behaviors of thousands, millions, or even billions of neurons.

Paracrine Signaling

Another kind of signaling that generates rhythms exists in the brain beyond the electrical signals we just examined.

Neurons and other cells (e.g., pinealocytes) send out signaling molecules that can diffuse throughout the body and brain. Melatonin is one such molecule. Serotonin and dopamine are often discussed for mood. I learned about oxytocin mainly from taking a breastfeeding class with my wife. How these signals are released as well as travel throughout the brain and body is an area of active study.

Paracrine signals are chemical signals from a neuron that are not isolated to a synapse. Once a neuron releases a paracrine signal, such as in response to an action potential within the cell, the paracrine signaling molecule can diffuse widely, both in the space between neurons in the brain and even into the blood to other regions of the body. Synaptic signals like GABA, which we think of as isolated to synapses, can "spill over" into the space beyond the synapse and affect the behavior of other neurons not near the signal. Paracrine signals are much slower than the electrical signals we mentioned earlier since signals slowly accumulate in the space between neurons and are sometimes called tonic because of this.

Paracrine signals can act in two different ways. They can directly change the voltage of a cell. We call such factors ionotropic. Or they can bind directly to the channels that allow currents to flow within the cell. The second kind of paracrine signal is where the paracrine factor binds and affects the cell's inner workings. This is called a metabotropic signal, which can affect the cell's metabolism and do other things like turn on genes. Metabotropic signaling can occur on much longer timescales than ionotropic signals. For example, it might signal to turn on a gene, which can then take hours to make a fully formed protein from the gene's instructions.

The spatial extent of paracrine signaling depends on many factors. First is the diffusion rate at which they travel. Faster diffusion means that signals will travel farther. Additionally, paracrine signals degrade. If they degrade quickly, they will stay within a smaller volume. Degradation is also carefully regulated by glial cells, which support neurons and control the environment neurons find themselves in. Glia uptake these signals and degrade them. If the glia does not like a signal, they can take it up, limiting the spatial extent to which it will signal. Moreover, a recently discovered glyphatic system clears fluid between neurons and all the signaling molecules.

Paracrine signaling can create waves of activity as the signal spreads. But likely, paracrine signals do not act alone. For instance, a single cell's signaling could start a wave, including specific currents that might bring a cell to a state where it releases a paracrine factor. That paracrine factor would then diffuse and cause some other neurons to signal. These neurons could then send excitatory or inhibitory signals across the brain through white matter connections. Figuring all of this out is going to take more than just clever thinking. It will require rigorous mathematical models containing all factors, perhaps even with millions of neurons, and then new mathematical techniques to understand how all of these factors generate rhythms.

New Methods to Simulate and Understand Rhythms of the Brain

There are many approaches to understanding rhythms of the brain at scale. They can show how rhythms in the brain can be generated as a single-cell, small network, or even whole-brain level. Other books, such as my own, published by the MIT Press, provide more mathematical detail, but at a high level, the reader can understand each of the approaches along with their appeal and limitations. The future of understanding the EEG and other whole brain phenomena relies on our ability to further these methods. As we can simulate the brain, we can test hypotheses for what is generating specific EEG behaviors.

At the most abstract level, we use ML to understand the brain's responses. The clear benefit of this approach is that ML is extremely good at working with and extracting information from large datasets. If one has thousands or more measurements, ML is a great approach. Even if one has just a few EEG recordings, they can be lengthy (e.g., hours) and contain fast signals (e.g., on the second timescale)—an enormous dataset. Thus long-term EEG recordings can be considered hundreds or thousands of experiments played one after another.

Moreover, an EEG can present data from many (up to sixty-four or more) channels, adding much data to the dataset. ML approaches are complex because they take a black box approach, as discussed in chapter 1. ML models can be suitable for prediction in some cases, but look inside them, and there typically is little insight to be gained. ML is built on what are dubbed "neurons," but are mathematical equations showing only the simplest of behaviors. These so-called neurons are not complex enough to show an action potential or any of the signals we have just mentioned. Signaling in neurons in the brain is much richer and more complicated than that used in ML. But perhaps future ML will utilize more complex descriptions of neurons.

A slightly less abstract way to analyze systems in the brain is to use firing rate models. These assume that each brain region can be understood by an average "firing rate," where the output is a simple, predetermined input function. Firing rate models are a little different from sigmoid models in NL. Nevertheless, researchers who use firing rate models can specify connections between specific brain regions based on physiological data rather than having the computer guess.

Another way to understand EEG experiments is to represent the behavior of a brain region probabilistically. This is called the population density approach since we simulate a probability that each neuron is in any particular state, which we call a *probability distribution*. This is a flexible approach and can allow neurons to show the many different types of behaviors we explored previously. One probability distribution could represent the state of millions of neurons in small brain regions, such as one of the hundred thousand voxels (regions) of the human brain imaged in an fMRI. Connecting them through synaptic or paracrine connections allows us to simulate whole-brain phenomena.

Finally, we can directly simulate each of the neurons and their connections rather than taking a regional approach like the population density one. This is the most accurate, but it also requires tremendous computing power. Games and cryptocurrencies make this possible (though only recently). As the demand for precise computer graphics increased, hardware companies like NVIDIA developed special hardware (graphics processing units [GPUs]) to do the mathematical calculations needed to render the graphics quickly. Crypto miners reprogrammed these cards to do the math to, for example, mine Bitcoin quickly. There is tremendous demand for computer games and crypto, creating vast market incentives for more powerful GPUs. For these reasons, GPUs have become fast and powerful computationally, and may allow for direct simulation of the brain. We have been able to do this for the mouse cortex, with its millions of neurons, but we are far from doing this for the human brain, with billions of neurons.

Deciphering Brain Signals

A fascinating age is dawning in understanding brain signals. As we have seen, individual neurons can generate many possible behaviors, and these behaviors become much more complex once neurons are connected in large networks. Mathematical and computational techniques have rapidly risen so that we can now, for instance, simulate the underlying physiological systems creating the EEG signals, such as the firing of individual neurons or paracrine signals in the cortex, and compare them to recorded signals. Fitting models to a recording is a tough challenge, taking thousands of simulations or more to understand the physiology of the EEG fully. Once this is done, though, it will usher in a new age of brain-computer interface. We might think about coffee, and our coffee maker will start to brew coffee. We might be too tired to drive, and a car's navigation system might kick in automatically. Differences in EEG signals might help us peer into the brain to discover diseases beyond those already diagnosed via an EEG, like epilepsy, and give early warning signs that cause individuals to seek help from their doctors.

We are in the infancy of a deep understanding of the physiology behind rhythms recorded by an EEG. It promises to hold the most information of any rhythms studied in the previous chapters. Because of the many types of signaling in the brain, it is also the most complex language to decode. Furthermore, it brings up ethical questions. Detecting fatigue in someone running a nuclear power plant or flying an aircraft seems a good choice, considering the potential risks involved. Yet detecting fatigue in my classroom students or even their thoughts in response to my lecture crosses a line.

METABOLISM

The devastating effects of metabolic disorders have touched many families. Type 2 diabetes is a national pandemic. My grandfather's health was greatly affected by diabetes, eventually leading to the loss of toes and a kidney. Not only does it impact an individual's health, diabetes burdens caretakers who work hard to manage its effects. The genetic predisposition to diabetes also burdens families as it can spread from one generation to the next.

The popular literature hones in on our high-sugar diet as the reason for the type 2 diabetes pandemic. While we typically focus on counting calories, rhythms are emerging as a significant factor. Our health depends on what we eat and when we eat it. I realized this when an ad appeared on my screen with a picture of someone my age who looked much healthier than I (frankly healthier than anyone my age might look). I could not resist the temptation to click and discover the secret of their success—perhaps my future success. If I looked like that, my students would pay more attention to my lectures. What came up was an advertisement for time-restricted eating, an idea with support from the circadian rhythms community where food intake is restricted to specific times of day. Rather than just being a fad, although some claims in the popular media are inaccurate, there *is* science to back up the assertions that controlling which times of day we eat, independent of the total number of calories, can improve health.

Understanding metabolic rhythms requires looking at the basic science of circadian rhythms. We will then discuss how daily rhythmic patterns affect metabolism. Metabolism, mainly glucose levels, can be measured by wearables. This presents great hope for keeping the blood sugar of diabetics in check. But few understand the language of metabolic rhythms, similar to that found in other rhythms in the body. Monitoring metabolic rhythms is at the forefront of wearable technology, and many innovations lie ahead; human metabolic rhythms are just beginning to be understood.

Going Back to the Beginning

To understand human metabolic rhythms, it is helpful to understand some concepts from metabolism in basic

Our health depends on what we eat and when we eat it.

organisms. Returning to the basics and studying these rhythms in simpler organisms is helpful. The continuous monitoring of metabolism was historically one of the oldest physiological rhythms studied. Metabolic rhythms have been observed in yeast metabolism since the 1950s. They were some of the first rhythms studied through mathematical models as well. Glycolysis is the process by which glucose is used and converted to energy stored in adenosine triphosphate, which acts as the battery of cells and provides the power for many cellular functions. Under certain conditions, glycolysis rhythms emerge within a period of minutes. These oscillations initially measured in yeast have been the subject of many mathematical models.[1]

Yet circadian and metabolic rhythms have long been linked. Circadian clocks may have evolved from metabolic rhythms. Cells in organisms so simple that sleep rhythms, melatonin, temperature, mood, or heart rate rhythms are not present still have circadian rhythms. They need these rhythms to determine when they are ready to produce energy or conserve it. Direct links between circadian and metabolic rhythms have been found. For example, the sirtuin proteins, which control glycolysis, have their expression controlled by the molecular circadian clock within cells. Metabolic rhythms are ubiquitous, and almost all cells in nature show circadian rhythms in metabolism.

We still don't know the details of how circadian and metabolic rhythms are linked. One of the most interesting

Metabolic rhythms are ubiquitous, and almost all cells in nature show circadian rhythms in metabolism.

examples is a simple unicellular organism, Acetabularia, an algae with unique properties. It has just one cell, but is one of the largest single-celled organisms in nature, growing to several centimeters in length. Twenty-four-hour rhythms in metabolism (particularly oxygen consumption) have been recorded in these organisms. Now remove the nucleus of a cell of Acetabularia, or even red blood cells in humans, which often do not have a nucleus. Since the molecular clock in animals is genetic, with the turning on of genes being part of the core feedback loops that generate timekeeping, one would expect not to see any rhythms. Amazingly, the twenty-four-hour rhythms of metabolism persist. So daily rhythms in metabolism are so crucial to this organism that you can remove any possibility of a traditional circadian clock, which regulates gene activation, at least as it functions in mammals, and rhythms persist. Surprisingly, there also are circadian rhythms in red blood cells, which lack much of the key mechanisms of circadian timekeeping.[2] How these cells generate these circadian rhythms remains a mystery.

There are other interesting aspects to how circadian and metabolic rhythms are intertwined. One example of this is a paper published twenty years ago based on metabolic rhythms in yeast.[3] They claimed that as the period of the circadian clock is lengthened or shortened, similar percentage increases in the rhythms of metabolic rhythms on the timescale of minutes were seen too. Moreover, circadian and metabolic rhythms control how quickly cells divide, which is the basis of cancer. How all of these rhythms are interlinked remains to be understood.

Circadian rhythms in metabolism make sense since many organisms only have food available at certain times. Simpler organisms may also need light for metabolism, which in the wild, is not available at night. This daily rhythm of the Earth has caused our body to regulate our metabolism and anticipate when increased metabolism may be needed. The modern lives of humans seem far from these concerns, but our bodies have ancient systems and structures that keep metabolic timing key.

The Timing of Food

Before modern times, humans likely consumed food during short intervals each day. Night times were not a time to eat for several reasons. First, without refrigeration or other ways to preserve food, it was impossible to store food so that humans could consume it at any time without risk. Second, nights were frequently a time for a community to huddle together for social reasons and because of the risk from nocturnal predators. Thus the body evolved to have food presented at certain times of the day. Food collection (e.g., hunting) and preparation (without modern appliances) were much more difficult. Food supplies were not as plentiful as they are currently in high-income countries, which many readers of these words are lucky to live in. Yet even in the wealthiest communities, some are still food insecure.

But with scarce food, the body needed ways to be ready whenever food was available. What if a flock of birds flew by around sunrise? What if the primary source of protein was a nocturnal rodent? What if the salmon entered the river at dusk? First, the body can jump-start its system to be prepared whenever food is presented. The body has little ways to understand unconsciously how big a meal is. So the system turns on from the first bite and gets ready. This is a better option than having food sit undigested in the body. The amount of food needed to jump-start these rhythms varies from person to person, based on their prior history of eating, size, and overall health; digestion is something we cannot do partially. Figure 8 shows the changes in glucose rhythms after meals.

The second factor is the circadian timing of meals. You may have experienced this in a similar way to me. Sometimes I wake up in the middle of the night. Biologically, this probably is not as bad as we might think it would be, at least for our ancient selves, since research suggests that before modern lighting, waking in bed during the night was much more common, especially as individuals were in bed for most of the night. In my modern world, however, these are lost hours, and with my alarm clocks (a.k.a. my

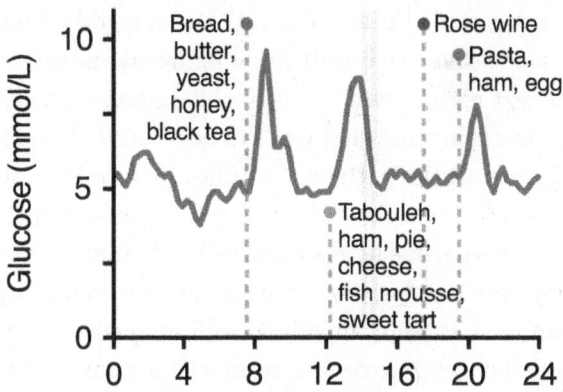

Figure 8 The change in glucose levels as measured by a continuous glucose monitor after meals. *Source:* Nicolas E. Phillips, Tinh-Hai Collet, and Felix Naef, "Uncovering Personalized Glucose Responses and Circadian Rhythms from Multiple Wearable Biosensors with Bayesian Dynamical Modeling," *Cell Reports Methods* 3 (2023): 100545.

kids) going off every morning, this comes at the expense of sleep. So I am desperate to go back to sleep as quickly as possible when I wake up in the middle of the night.

It has been several hours since I last ate, and I can feel my stomach empty (which is not real hunger considering I just ate and will eat again tomorrow), so I immediately head to the kitchen for a quick midnight snack. Not only does this turn on my digestive system, but the food sets the clocks in the liver. My ancient circadian timekeeping system wonders if there is some special plant that only blooms at night and will be eaten by other animals before dawn (or maybe it is the animal that I will eat). Anyway, this food shifts the liver clocks to be ready to eat in the middle of the night. The next night, I was up even a bit earlier, with my body ready to eat again. What was I to do? I need to sleep, so I eat again, and the pattern continues. Before long, it had been a week or two since I had a restful night's sleep.

These nocturnal meals do even more than shift the clocks in my liver. In experiments in rodents, the same phenomena are seen. Present an animal with food when they should be asleep over a couple twenty-four-hour cycles. Then stop. The animal will still wake up just in time to eat as in previous days, at the expense of sleep, even if no food is available. In animals, we can probe to see what has changed physiologically within the body. Although it is somewhat debated, there is data that a timekeeping region in the hypothalamus, separate from the SCN, is called into service. This region commands the whole body, not just the liver, to be ready for this food source. Scientists call this the food-entrainable oscillator. If you eat when you should sleep, the brain builds a new alarm clock in a region that typically does not keep time to wake you up in time for the next meal.

Time-Restricted Eating

For this reason, many scientists, particularly Satchin Panda in his book *The Circadian Code*, have suggested that individuals follow time-restricted eating.[4] Time-restricted eating limits when we consume foods, and are set at an eight, ten, or twelve-hour window rather than the sixteen-hour or more window many individuals follow. While healthy foods are encouraged, it is not a diet, nor are the benefits of fasting for more than sixteen hours widely accepted. In its most accepted form, calorie restriction is not required (though limiting calories in the modern diet can be helpful in some cases). The benefits of this are not only in the potential to lose weight but also the potential to sleep better and ensure stronger rhythms throughout the body, which could make us productive. According to experts, it is a simple step to improve health.

Still, more than restricting food to a total number of hours may be required. Eating is needed at the correct phases of the circadian rhythms of our metabolism. Take, for example, studies of night shift workers. It is well-documented that eating at the wrong circadian time can yield significant health risks. But what are the right and bad times of the day? This is where a new type of wearables can play an important role.

Metabolic Rhythms in Humans

Shift workers are especially at risk for metabolic disorders. In a key experiment, normal subjects lived a twenty-eight-hour day in carefully controlled laboratory conditions.[5] They did not increase their caloric intake. Yet as their circadian rhythms became misaligned from their sleep-wake cycles and factors like insulin were produced in the wrong quantities at the wrong times, as is typical in shift workers, they developed signs of diabetes. Shift workers have higher rates of diabetes than normal individuals. The particular mechanisms for this increased rate of diabetes still need to be determined, though.

Carefully controlled laboratory studies have also measured rhythms affecting metabolic health, such as cholesterol. These factors have strong circadian rhythms in normal individuals. But metabolic rhythms shift when individuals undergo shift work, even when that shift occurs in carefully controlled laboratory settings. What likely emerges is a mixture of two factors. First, eating meals at different times can cause responses that increase digestion elements. Additionally, metabolic rhythms are particularly sensitive to mealtimes. For instance, the circadian clocks of cells within the liver shift when meals are presented away from their normal times.

Glucose Rhythms

Millions of people track their glucose levels. There are several ways to measure glucose levels in the body. Traditionally, individuals prick their fingers to extract blood and measure glucose levels. Recent wearable technology allows individuals to use CGMs to keep an eye on glucose levels. Yet we can track metabolism in other ways. We can draw blood at different times and measure metabolites. As this data emerges, a new understanding of how the rhythms of metabolism work offers great potential health benefits.

We are not different from animals or yeast in our need to regulate glucose levels. Low glucose levels and the body cannot properly function. We can feel faint or worse. High glucose levels can lead to diabetes. For these reasons, individuals with diabetes must continually check their glucose levels. If it is too low, they must eat or drink foods that will raise glucose. If it is too high, individuals with diabetes need insulin. I remember my grandfather constantly pricking himself to test for glucose in the blood. There must be better ways.

Now we can continually monitor glucose with a CGM device. CGMs look much like patches and avoid the need for pricks. When the CGM is attached to the skin, a small probe is inserted underneath the skin. That probe can painlessly sense the amount of glucose within the body many times an hour. Data can then be uploaded to a smartphone via Bluetooth, and apps can analyze this data to determine if glucose is too low or high. This data contains many different factors, such as heart rate.

But what do these rhythms in glucose look like? The answer is similar to heart rate, temperature, and other rhythms we have studied. Meals will induce a spike in glucose levels, which will then diminish over a timescale of about an hour or two (see figure 8). This process is longer than the effects of meals on heart rate, which is closer to an hour or less, but has a similar effect. Additionally, there is a basal circadian rhythm in glucose levels, separate from when meals occur. Thus the body is not only responding to meals but instead anticipates them. How this circadian rhythm is regulated is still under investigation. Glucose has a different profile and timing, however, than melatonin or other rhythms. It also can change from individual to individual and day to day, similar to temperature and heart rate rhythms.

Like the other rhythms we have studied, glucose has its circadian language. While heart rate listens most to activity and melatonin listens most to light, glucose listens most to meals. There are indeed effects of activity, and perhaps light, similar to these other rhythms, but they are in fact smaller. Too much physical exertion can cause glucose levels to drop very low. Stress can affect glucose levels too. Sleep is an essential regulator of glucose as well. Many people with diabetes worry about glucose levels that are too low during sleep, just as blood pressure, heart rate, and other factors are low. Low glucose levels during sleep can cause a host of unpleasant and potentially dangerous side effects, including nightmares.

In this way, glucose regulation is like the regulation of many other biological rhythms studied in previous chapters. Much less is known about it, but this will soon change.

There has been a rapid expansion of CGMs, with millions of individuals continuously monitoring glucose levels. Using the knowledge of the other rhythms discussed previously will give us many hypotheses to test about glucose rhythms. Nevertheless, the language of glucose rhythms is uniquely tied to meals and has its own signatures.

Diabetes and Insulin

CGMs are just one part of the equation for diabetes care. If glucose levels are too low, an individual might need to ingest sugar quickly. If glucose levels are too high, insulin might need to be supplied. This can now be done automatically via wearables. Algorithms on the phone can analyze glucose measurements. After each measurement, they can decide how much insulin should be given to the body, and the insulin is automatically deployed. In this way, insulin is the most advanced rhythm in that many people are monitoring it with wearables and automatically changing it with algorithms.

The circadian control of insulin seems to follow the same paradigm mentioned for melatonin, heart rate, or other circadian rhythms. Internal molecular clocks control its release, which is also coordinated by signaling from the brain.[6] These algorithms do not account for circadian rhythms or many other factors, and simply keep glucose within a fixed range. A rhythmic approach may be important in fighting diabetes and lead to more efficient treatments. Furthermore, better links to physiology could help understand glucose dynamics. Glycolytic oscillations are part of the process of insulin secretion and similar in mechanism to those studied in yeast. Mathematical and conceptual models are needed to link different timescales from minutes (glycolysis) to days (circadian), requiring special mathematical and computational skills.

Breath

It is unlikely that individuals will want to give blood samples throughout the day to measure the many metabolic rhythms in the body. CGMs might eventually evolve to measure other metabolites. Yet they still involve a probe entering the skin. An exciting new way to test for metabolic rhythms is through breath. Many metabolites are volatile, meaning they can enter the air. Volatile compounds can be detected with techniques like gas chromatography. Some have begun to study these metabolic compounds in the real world. A group at the University of Zurich has shown that the levels of metabolites in breath are a good predictor of circadian rhythms, at least in healthy individuals. New research is now focusing on detecting fatigue through breath. These studies typically involve an individual breathing into a device or bag that collects the air to be later analyzed.

Xudong Fan, though, has taken this to another level. First, he developed portable devices to analyze breath and determine underlying compounds. Second, he developed a prototype wearable device worn like a patch on the skin. The skin constantly releases volatile compounds, and it is possible to measure metabolic rhythms from such devices. These offer a new generation of wearables that monitor many (tens, if not hundreds) of compounds in real time.

The Future of Understanding Metabolic Rhythms

Rhythms in metabolic function span many different timescales. Natural metabolic processes, like glycolysis, can biochemically generate rhythms on the order of minutes. These rhythms are seen in yeast and mammals, such as insulin generation. Rhythms are induced by meals, which could occur three times a day. These mealtimes are anticipated, and eating at the wrong time one day will yield shifts in the coming days. Circadian rhythms in sleep and heart rate also affect metabolism, and vice versa. Factors like sleep and activity play a role in metabolic rhythms too, but their effects should be smaller than that of meals.

Real-world glucose monitoring is already occurring. The vast amount of data we will get from the real world from CGMs and other devices will revolutionize the treatment of diabetes and other metabolic disorders. If large datasets from many individuals could be studied, many insights into glucose dynamics and how it varies between individuals could be discovered. But breath and skin wearables also offer the possibility of continuously collecting more data on many metabolic compounds. This new data is being generated quickly enough that a host of information and personalized treatments will be available in the next five to ten years.

We require more than just collecting the data. As previously described, mathematical techniques will be needed to piece apart rhythms on different timescales. Are there glucose rhythms on the order of minutes, like in yeast? Such rhythms may exist in some tissues, but are too hard to detect in a whole-body assay like blood, which contains many different cell types. We may also find metabolic rhythms on the timescale of minutes, as have been seen in some other animals. Complicating this is that blood or other samples are heterogeneous collections of different types of cells whose rhythms may not be synchronized.

Understanding metabolic rhythms in the real world requires much more work. Much of what we know comes from human experiments that have occurred in the lab, where meals are replaced by regular isocaloric snacks. Again, a balance of real-world and carefully controlled laboratory data is needed to fully understand the rhythmic language of metabolism.

PUTTING IT ALL TOGETHER

New, noninvasive ways to measure physiology over time will be developed with each passing year. As more physiological data becomes available, we will learn more about our physiology and how it adapts to the modern world. This will bridge the conscious and unconscious worlds of our body. Wearables will detect disease before we feel symptoms. Our meals, activities, and sleep will be planned for optimal health and peak performance. This information will be presented to us, and we can decide when or when not to follow it.

The previous chapters have equipped you with many of the necessary tools and frameworks to understand the rhythms of your body. While each rhythm is unique, we find that they all have common features by carefully considering these many rhythms. They are all affected by sleep and activity. There are underlying circadian rhythms to all physiological signals; the body constantly predicts when a physiological system is most or least helpful, when it should be on high alert or can take a snooze. Key to deciphering these signals are timescales. A brief increase in heart rate because of a sprint is shorter than an increase in heart rate due to adrenaline, which is different than the effects of sleep, which are shorter than the twenty-four-hour rhythm. Algorithms can decipher these different rhythms.

Studying each rhythm teaches us more about what each physiological system tries to do. Its instructions in DNA were crafted over millennia to adapt our bodies to the natural world best. Modern lifestyles have drastically changed, however, even in the past ten years. This discord between what our bodies were designed to do and what the modern world expects of them may be part of the cause of many diseases, from cancer to depression. Understanding this discord allows us to treat disease in new ways and live happier lives.

We have the sensors to achieve this utopian future. What is needed are algorithms and an understanding of how human physiology works in the real world. You can participate in this future right now. Check your smartphone for the data it has already collected. Note special circumstances: getting married, being diagnosed with a disease, or even being the victim of road rage, and checking how your body responds. But most signals are hidden from the naked eye. Only through the microscope of mathematics can we understand what the body is saying.

You can donate your data to researchers, such as through our research group's Social Rhythms app, and we can use your information for science and provide you with circadian information as well as *digital twins* as discussed below. These apps can also translate your data into specific responses to our environments. But this is just the beginning. The most exciting advances in understanding the systems of our body will come from mathematics. Here are some of the most promising avenues for future advances.

Digital Twins

With large-scale datasets on human physiological systems, from fMRI images or animal studies where we can visualize every cell in a tissue, we can now form a new generation of mathematical models called digital twins. Digital twins are detailed mathematical models that aim almost perfectly to replicate a specific system and are specific to an individual. So while we might have a general mathematical model, like for bipolar disorder, as discussed in chapter 6, a digital twin would have a detailed accounting for all the physiological processes underlying bipolar and have parameters tuned to an individual. These models have tremendous potential for realism and can carefully match data from a single human. If coded efficiently, they can run close to real time too. Surgeons are developing digital twins to help them precisely operate on tumors. Similarly, digital twins can be conceived of based on the physiological processes generating vital signs, which could give an extremely accurate understanding of the rhythms in your body and personalized predictions.

What makes this possible is a law about computing. Moore's law states that a computer's number of transistors (the most critical circuit element) can double about every two years. This kind of growth is challenging to appreciate fully. At the moment, you can buy a GPU with over eighty billion transistors for a cost that many can afford. Unfortunately, we will need to look to other ways to continue this growth, like quantum computers, as Moore's law is hitting a plateau based on some physical limits for how small a transistor can be. Yet few applications efficiently use even a fraction of the eighty billion transistors at our disposal on a GPU.

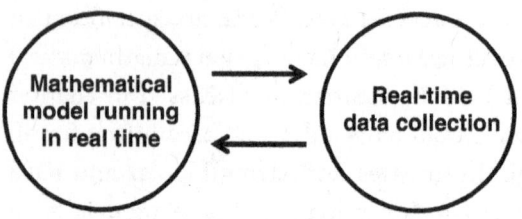

Figure 9 Schematic of a digital twin.

Digital twins can be conceived of based on the physiological processes generating vital signs, which could give an extremely accurate understanding of the rhythms in your body and personalized predictions.

Digital twins can run simultaneously on physiological systems and give alerts when they detect abnormal states. They carefully integrate measurements about the system, and merge this with prior knowledge of how the system works to better understand its current state and predict what is happening "under the hood." The first examples go back to a mathematical technique called the *Kalman filter*.

Understanding Time Series

When we measure rhythms, we make many measurements. Critical to understanding rhythms is to determine if the measurements, taken as a whole, give more information than each data point. Rudolfo Kalman determined one method to do this. He wondered, given a model we can trust (a big assumption, which we will return to), can new measurements be best integrated with the model to estimate the state of the system better? How could prior measurements in a time series, such as previous heart rate measurements, help us to predict the system's current state beyond the latest measurement?

On a train ride, Kalman realized that past measurements could also be used if the model could translate the system's behavior in the past to a prediction of the system's current state. Prior system measurements could be translated into a measurement of the current system state, with some prediction error. The prediction error typically increases as time goes back, but if we have many successive measurements of a system, why not use them all (along with a model) to get as accurate a measurement of the current state as possible?

Then came some mathematics, which was not trivial but instead the kind that someone with an undergraduate engineering degree could understand. Given prior measurements, Kalman derived a formula for the most accurate way to update a system's state. He then proved that no other simple method could do this better. This is how the Kalman filter was born.

Kalman filtering has directly impacted your life in many ways. Take GPS, for example. GPS works because the Kalman filter works. Now GPS is super quick, but until recently (and if you try GPS in the woods on a cloudy day), you will wait while the device is "locating" you. What is happening is that many measurements are being taken, and these are being integrated with the current measurements using the Kalman filter to give a more precise location. After all of this, you have an amazingly accurate picture of where you are on the vast planet.

We are building a kind of GPS for your body with this framework in Michigan's Social Rhythms app. Wearables will continually take measurements of physiological systems. Detailed mathematical models, perhaps down to individual cells, will then be simulated to integrate past measurements with the current one, giving a more accurate picture of the system's current state. *Anomaly detection* will then check to see if the system is pathological. If so, personalized predictions will be presented to you so you can right the course. Like GPS, these predictions will be personalized and in real time. If you cannot follow these predictions, you can do whatever is best for you, and the system will "recalculate" and offer a different set of predictions.

One example of how we did this was with our group's prior Entrain app for travelers. Individuals downloaded the app before traveling across time zones. The app offered them suggestions, such as when to avoid light when they traveled to avoid jet lag, based on a mathematical model. There are times when you just cannot avoid sunlight. For instance, suppose you fly to Tokyo to give a business presentation and it turns out the business presentation is at a time when you should avoid light. In that case, you should give your business presentation. The free Entrain app recalculates and gives you new recommendations. It was used by hundreds of thousands of travelers.

What was particularly interesting to us was how individuals used these suggestions to shift their circadian rhythms in the real world. We knew that individuals would follow them differently because the constraints on doctors, students, or businesspeople are quite different. For this reason, much of the mathematical work to build Entrain's computational engine focused on giving individuals flexible rather than "static" recommendations. When we analyzed the data, we found that when individuals aligned their schedules closer to what we suggested, they avoided some of the effects of jet lag. This proved the principle of digital twins for wearable data, albeit in a small case.

This is just the start. Much more detailed models of the circadian system could have been used. We could offer better algorithms and suggestions. Rather than avoid jet lag, apps could be designed to help shift workers or many other scenarios. With sleep tracking, digital twins could make optimal recommendations for how to get deeper sleep. Digital twins and recommendations for the best exercise for athletes could track heart rate. We could track mood and interventions taken before mood episodes. This technology will come as soon as we figure out the relevant mathematics. Moreover, our digital twins will incorporate much more information from ever-growing wearable sensors. Models will also get more extensive and more accurate. GPS for health will become integrated into our daily lives.

Privacy and Ethics

We discussed privacy briefly in chapter 1. Having explored many of the frameworks that can be used to understand rhythms and insights gained from rhythmic data, returning to this important topic is helpful. At face value, the data we examined in earlier chapters may seem innocuous. Knowing someone's heart rate was eighty-seven at some point is low-risk knowledge. Yet outlier measurements could indicate that a person could have diabetes, a heart condition, or an affective disorder. Combining the data could yield other insights that an individual would want to keep private.

For these reasons, our group typically follows these guidelines: individuals should fully consent to sharing the data, all data we receive is anonymous, individuals can freely stop sharing data and also request to have the data we have collected from them removed from our servers, we typically do not make raw data publicly available, and individuals who share data with us should get some new insights about their data. Some studies have special restrictions, but frequently, with more work, we can meet these guidelines. For example, we had to build special information technology systems to send users insights about their data while not knowing who they are.

Another primary concern is that companies (rather than research groups) with your data can often sell this data without your permission or even knowledge. Many companies have this as a central part of their business plan. Companies do this without keeping you anonymous—a practice that is dangerous at face value. Could a future prospective employer pay a company for this data on you and use it to screen you as a job applicant without your permission? I am not a lawyer or member of government, but it does seem an area that may benefit from more consumer protections.

Genetics

A grandmother approached a sleep clinic in Utah asking for help. Many other clinics had turned her down. She would wake up early every morning and feel tired in the early evening, often going to bed before others might have dinner. Moreover, she saw the same behavior in her grandchildren. While many might believe this was a lifestyle choice, she felt it was physiological. Most important, she did not want her grandchildren to live with the same difficulties and stigmas she faced. The physician listened and recruited a team of experts, including Louis Ptacek, an expert in medical genetics, and David Virshup, a biochemist and oncologist, to figure out what was happening.

Circadian rhythms are famously genetic. This grandmother had one bit of her genetic code altered in the regions that code for a protein involved in intracellular timekeeping. This caused the clock to have a short internal period, which as discussed in chapter 3, caused her to wake early when she tried to live a twenty-four-hour day. This is a case of familial advanced sleep phase syndrome in less than 1 percent of the population.

Genetics affects any of the rhythms we looked at in prior chapters. Bipolar disorder has a high chance of being passed down from one generation to the next, as do metabolic disorders. Ptacek and his collaborator Ying Hui Fu continue to study families that sleep at unusual times as well as individuals who seem to get less than 6.5 hours of sleep a night and see no deficits for the lack of sleep. Ideally, the framework to understand your rhythms could be personalized with a genetic test. This is the field's holy grail, especially considering how large and complex genetic datasets are. But if we could figure out which changes in the genetic code change personalized parameters of rhythms, our ability to predict and understand rhythms would be vastly improved. Stay tuned!

Health Disparities

The most important impact of tracking and analyzing these physiological rhythms will be alleviating health disparities. Doctors are in short supply, even in the most resourced communities. In low-income countries or underresourced communities, the number of doctors is shockingly low, perhaps with one specialist for every hundred thousand people. The only way to make up for this, at scale, is with wearable technology, which can be deployed inexpensively. Several biomedical foundations have annual budgets that can afford to give a standard wearable to every citizen of a typical low-income country. With the proper infrastructure, the data from the wearables could be sent to secure servers, where analysis would be performed and optimal recommendations given. An individual may be warned about the possibility of an upcoming mood episode or perhaps they are warned about a possible COVID infection. Maybe they are told that their circadian clock is misaligned.

At scale, these interventions could provide needed first-round care for millions of people. Moreover, they might be able to triage patients to determine who needs scarce resources the most. One could see who is responding best to treatments and who the algorithms are not able to classify well. One could also determine who is doing the best at following the recommendations and who is ignoring them the most.

It will take teams of researchers to figure out these solutions. With such limited medical resources, how can we reasonably deploy resources? Such decisions cannot be made by algorithms alone. In addition, we must understand which interventions individuals are most likely to follow. Wasting resources on interventions that could work but that individuals need to follow through on will not help. Teams of medical doctors, mathematicians, ethicists, engineers, human-machine interface experts, and programmers are needed to welcome this new age of rhythmic self-discovery.

ACKNOWLEDGMENTS

I extend special thanks to the MIT Press team, especially Bob Prior, Anne-Marie Bono, and Caroline Helms. I will always be grateful for Bob's expert hand in guiding me through the rough waters of publishing. Thanks to many in the field who read through drafts of this manuscript to offer comments, including the students of Math 463, Math 564, and the SRBR trainee day participants, particularly William Schwartz and Delaney Beckner. Finally, this manuscript was written anywhere besides my office: Claryville, Condado, downtown Detroit, Nairobi, and anywhere between.

Adrenaline
A hormone that can indicate stress and controls heart rate.

Anomaly detection
A class of methods in the ML literature that determine if an unusual or unexpected value occurs.

Antidromic
Proceeding in the opposite direction of what is intended.

Basal heart rate (BHR)
Heart rate when other factors (sleep, activity, or stress) have been controlled.

Circadian amplitude
The strength of the signal from a circadian clock.

Circadian (internal) period
The period of the rhythms of the daily timekeeping systems if all input factors were removed. Surprisingly, this is not precisely twenty-four hours.

Circadian phase
This is the time of day your body (or a part of your body) thinks it is.

Circadian rhythm
A rhythm of approximately a day. There are circadian rhythms in heart rate (CRHR), melatonin, temperature, mood, and other vital signs. They are typically controlled by intracellular clocks that tick at a rhythm of about a day.

Crepuscular
Occurring twice a day, such as at dusk and dawn.

Digital twins
Realistic mathematical models are run side by side with data collection to give more accurate predictions of a (physiological) system.

Entrainment
The ability of a clock to match its period to that of a rhythm signal.

Glymphatic system
A system that clears waste from the brain.

Heart rate correlation
How much a heart rate measurement depends on the next. This dependency could reflect the dynamics of hormones such as cortisol, cleared on an hour's timescale.

Hemicircadian
A period of about twelve hours

Hypothalamus
A region of the brain that controls hormone release and many systems in the body.

Internal Heart Rate Variation
The variation of the heart rate measurements from one minute to the next. Thus it measures variations in internal signals that control heart rate.

Kalman filter
An algorithm uses a mathematical model and time series data to estimate the state of a system.

Melatonin
Melatonin is a hormone with a strong circadian rhythm that indicates whether the body thinks it is night or day. Dim light melatonin onset is a technique for determining a phase from a melatonin rhythm.

Mixed episodes
Mood episodes that contain features of depression and mania.

Mood variability
This measures the overall strength of factors that influence mood, such as internal variation coming from the stochasticity of neuronal processes and external events that could affect mood.

Neuroticism speed
A parameter in a mood model that determines how quickly mood returns to its baseline state after a perturbation.

Phase response curve
A characterization of the effects of a particular signal (e.g., light or activity) on a (circadian) clock. Signals that speed the clock are called phase advances. Signals that slow the clock down are called phase delays.

Postural orthostatic tachycardia syndrome (POTS)
A short-term, unusually high increase in heart rate when posture changes quickly

Probability distribution
A mathematical statement of the probability of even occurring.

Sleep drive
The ability of an individual to fall asleep.

Sleep stage
A measure of the depth of sleep. REM or NREM, sleep, or deep and light (respectively) sleep. Humans usually cycle between these sleep stages every forty-five minutes or so. Polysomnography is the gold standard way to determine sleep stage.

Synapses
A region of signaling between two neurons.

Synaptic homeostasis hypothesis
A theory is that sleep occurs to compensate for the strengthening of synapses in the brain during the waking day.

NOTES

Chapter 1

1. William Gilpin, Yitong Huang, and Daniel B. Forger, "Learning Dynamics from Large Biological Data Sets: Machine Learning Meets Systems Biology," *Current Opinion in Systems Biology* 22 (2020): 1–7.

2. John D. Kelleher, *Deep Learning* (MIT Press, 2019).

3. Xiaoman Xing, Zhimin Ma, Mingyou Zhang, et al., "An Unobtrusive and Calibration-Free Blood Pressure Estimation Method Using Photoplethysmography and Biometrics," *Scientific Reports* 9 (2019): 8611.

4. Colleen M. Bartman, Yoshimasa Oyama, and Tobias Eckle, "Daytime Variations in Perioperative Myocardial Injury," *Lancet* 391 (2018): 2104; Marc D. Ruben, David F. Smith, Garret A. FitzGerald, and John B. Hogenesch, "Dosing Time Matters," *Science* 365 (2019): 547–549.

5. Panagiotis Fotiadis and Daniel B. Forger, "Modeling the Effects of the Circadian Clock on Cardiac Electrophysiology," *Journal of Biological Rhythms* 28 (2013): 69–78.

6. Merrill M. Mitler, Mary A. Carskadon, Charles A. Czeisler, et al., "Catastrophes, Sleep, and Public Policy: Consensus Report," *Sleep* 11 (1988): 100–109.

7. Emi Nagoshi, Camille Saini, Christoph Bauer, et al., "Circadian Gene Expression in Individual Fibroblasts: Cell-Autonomous and Self-Sustained Oscillators Pass Time to Daughter Cells," *Cell* 119 (2004): 693–705.

8. Dae Wook Kim, Ja Min Byun, Jeong-Ok Lee, et al., "Chemotherapy Delivery Time Affects Treatment Outcomes of Female Patients with Diffuse Large B Cell Lymphoma," *JCI Insight* 8 (2023): e164767.

Chapter 2

1. Richard Kronauer, personal communication with the author, 1999.

2. Giulio Tononi and Chiara Cirelli, "Sleep Function and Synaptic Homeostasis," *Sleep Medicine Review* 10 (2006): 49–62.

3. Brendon O. Watson, Daniel Levenstein, J. Palmer Greene, et al., "Network Homeostasis and State Dynamics of Neocortical Sleep," *Neuron* 90 (2016): 839–852.

4. J. Christopher Ehlen, Allison J. Brager, Julie Baggs, et al., "Bmal1 Function in Skeletal Muscle Regulates Sleep," *eLife* 6 (2017): e26557.

5. Dag Stenberg, Erik Litonius, Linda Halldner, et al., "Sleep and Its Homeostatic Regulation in Mice Lacking the Adenosine A1 Receptor," *Journal Sleep Research* 12 (2003): 283–290.

6. Daniel B. Forger, *Biological Clocks, Rhythms and Oscillations: The Theory of Biological Timekeeping* (MIT Press, 2017).

7. Benjamin Collins, Sara Pierre-Ferrer, Christine Muheim, et al., "Circadian VIPergic Neurons of the Suprachiasmatic Nuclei Sculpt the Sleep-Wake Cycle," *Neuron* 108 (2020): 486–499.

8. Forger, *Biological Clocks, Rhythms and Oscillations*.

9. Olivia J. Walch, Amy Cochran, and Daniel B. Forger, "A Global Quantification of 'Normal' Sleep Schedules Using Smartphone Data," *Science Advances* 2 (2016): e1501705.

Chapter 3

1. Lorraine Potocki, Daniel Glaze, Dun-Xian Tan, et al., "Circadian Rhythm Abnormalities of Melatonin in Smith-Magenis Syndrome," *Journal of Medical Genetics* 37 (2000): 428–433.

2. Alfred J. Lewy and Robert L. Sack, "The Dim Light Melatonin Onset as a Marker for Circadian Phase Position," *Chronobiology International* 6 (1989): 93–102.

3. Garen V. Vartanian, Benjamin Y. Li, Andrew P. Chervenak, et al., "Melatonin Suppression by Light in Humans Is More Sensitive than Previously Reported," *Journal of Biological Rhythms* 30 (2015): 351–354.

4. Martin Zurl, Birgit Poehn, Dirk Rieger, et al., "Two Light Sensors Decode Moonlight Versus Sunlight to Adjust a Plastic Circadian/Circalunidian Clock to Moon Phase," *Proceedings of the National Academy of Sciences of the United States of America* 119 (2022): e2115725119.

5. Ellen R. Stothard, Andrew W. McHill, Christopher M. Depner, et al., "Circadian Entrainment to the Natural Light-Dark Cycle Across Seasons and the Weekend," *Current Biology* 27 (2017): 508–513.

6. Kurt Kräuchi, Christian Cajochen, Esther Werth, and Anna Wirz-Justice, "Alteration of Internal Circadian Phase Relationships After Morning Versus Evening Carbohydrate-Rich Meals in Humans," *Journal of Biological Rhythms* 17 (2002): 364–376.

7. Raymond P. Najjar and Jamie M. Zeitzer, "Temporal Integration of Light Flashes by the Human Circadian System," *Journal of Clinical Investigation* 126 (2016): 938–947.

8. Daniel B. Forger, Megan E. Jewett, and Richard E. Kronauer, "A Simpler Model of the Human Circadian Pacemaker," *Journal of Biological Rhythms* 14 (1999): 532–537.

9. Kirill Serkh and Daniel B. Forger, "Optimal Schedules of Light Exposure for Rapidly Correcting Circadian Misalignment," *PLOS Computational Biology* 10 (2014): e1003523.

10. Tiecheng Liu and Jimo Borjigin, "Reentrainment of the Circadian Pacemaker Through Three Distinct Stages," *Journal of Biological Rhythms* 20 (2005): 441–450.

Chapter 4

1. Nathaniel Kleitman, *Sleep and Wakefulness as Alternating Phases in the Cycle of Existence* (University of Chicago Press, 1939).

2. Christopher Flora, Jonathan Tyler, Caleb Mayer, et al., "High-Frequency Temperature Monitoring for Early Detection of Febrile Adverse Events in Patients with Cancer," *Cancer Cell* 39 (2021): 1167–1168; Dae Wook Kim, Ja Min Byun, Jeong-Ok Lee, et al., "Chemotherapy Delivery Time Affects Treatment Outcomes of Female Patients with Diffuse Large B Cell Lymphoma," *JCI Insight* 8 (2023): e164767.

3. Annabelle Ballesta, Pasquale F. Innominato, Robert Dallmann, et al., "Systems Chronotherapeutics," *Pharmacology Review* 69 (2017): 161–199.

4. Richard E. Kronauer and Megan E. Jewett, "The Relationship Between Circadian and Hemicircadian Components of Human Endogenous Temperature Rhythms," *Journal of Sleep Research* 1 (1992): 88–92.

5. Kronauer and Jewett, "The Relationship Between Circadian and Hemicircadian Components."

6. Arthur T. Winfree, *The Geometry of Biological Time* (Springer-Verlag, 1980).

7. Emery N. Brown and Harry Luithardt, "Statistical Model Building and Model Criticism for Human Circadian Data," *Journal of Biological Rhythms* 14 (1999): 609–616.

8. Ethan D. Buhr, Seung-Hee Yoo, and Joseph S. Takahashi, "Temperature as a Universal Resetting Cue for Mammalian Circadian Oscillators," *Science* 330 (2010): 379–385.

9. Brown and Luithardt, "Statistical Model Building."

10. Cliff B. Saper, Thomas E. Scannell, and Jun Lu, "Hypothalamic Regulation of Sleep and Circadian Rhythms," *Nature* 437 (2005): 1257–1263; Steven A. Brown, Gottlieb Zumbrunn, Fabienne Fleury-Olela, et al., "Rhythms of Mammalian Body Temperature Can Sustain Peripheral Oscillators," *Current Biology* 12 (2002): 1574–1583.

11. Colin S. Pittendrigh, "On Temperature Independence in the Clock System Controlling Emergence Time in Drosophila," *Proceedings of the National Academy of Sciences of the United States of America* 40 (1954): 1018–1029.

12. Min Zhou, Jae Kyoung, Gracie Wee Ling Eng, et al., "A PERIOD2 Phosphoswitch Regulates and Temperature Compensates Circadian Period," *Molecular Cell* 60 (2015): 77–88.

Chapter 5

1. Clark Bowman, Yitong Huang, Olivia J. Walch, et al., "Characterizing the Human Heart Rate Circadian Pacemaker through Widely Available Wearable Devices," *Cell Reports Methods* 1 (2021): 100058.

2. Jai Eun An, Kyung Ho Kim, Seon Joo Park, et al., "Wearable Cortisol Aptasensor for Simple and Rapid Real-Time Monitoring," *ACS Sensors* 7 (2022): 99–108.

3. Caleb Mayer, Jonathan Tyler, Yu Fang, et al., "Consumer-Grade Wearables Identify Changes in Multiple Physiological Systems during COVID-19 Disease Progression," *Cell Reports Medicine* 3 (2022): 100601.

4. Nathaniel Kleitman and Esther Kleitman, "Effect of Non-Twenty-Four-Hour Routines of Living on Oral Temperature and Heart Rate," *Journal of Applied Physiology* 6 (1953): 283–291.

5. Philippe Boudreau, Wei-Hsien Yeh, Guy A. Dumont, and Diane B. Boivin, "Circadian Variation of Heart Rate Variability Across Sleep Stages," *Sleep* 36 (2013): 1919–1928.

6. Yitong Huang, Caleb Mayer, Olivia J. Walch, et al., "Distinct Circadian Assessments from Wearable Data Reveal Social Distancing Promoted Internal Desynchrony Between Circadian Markers," *Frontiers in Digital Health* 3 (2021): 727504.

7. Huang et al., "Distinct Circadian Assessments."

8. Mayer et al., "Consumer-Grade Wearables."

Chapter 6

1. Kay R. Jamison, *An Unquiet Mind* (A. A. Knopf, 1995).

2. Amy L. Cochran, André Schultz, Melvin G. McInnis, and Daniel B. Forger, "Testing Frameworks for Personalizing Bipolar Disorder," *Translational Psychiatry* 8 (2018): 36.

3. Cochran et al., "Testing Frameworks."

4. Cochran et al., "Testing Frameworks."

5. Amy L. Cochran, Melvin G. McInnis, and Daniel B. Forger, "Data-Driven Classification of Bipolar I Disorder from Longitudinal Course of Mood," *Translational Psychiatry* 6 (2016): e912.

6. Diane B. Boivin, Charles A. Czeisler, Derk-Jan Dijk, et al., "Complex Interaction of the Sleep-Wake Cycle and Circadian Phase Modulates Mood in Healthy Subjects," *Archives of General Psychiatry* 54 (1997): 145–152.

7. Benjamin Shapiro, Yu Fang, Srijan Sen, and Daniel Forger, "Unraveling the Interplay of Circadian Rhythm and Sleep Deprivation on Mood: A Real-World Study on First-Year Physicians," *PLOS Digital Health* 3 (2024): e0000439.

8. Yu Fang, Daniel B. Forger, Elena Frank, et al., "Day-to-Day Variability in Sleep Parameters and Depression Risk: A Prospective Cohort Study of Training Physicians," *NPJ Digital Medicine* 4 (2021): 28.

9. Minki P. Lee, Dae Wook Kim, Yu Fang, Ruby Kim, Amy S. B. Bohnert, Srijan Sen, and Daniel B. Forger, "The Real-World Association Between Digital Markers of Circadian Disruption and Mental Health Risks," *npj Digital Medicine* 7 (2024), https://www.nature.com/articles/s41746-024-01348-6; Dongju Lim, Jaegwon Jeong, Yun Min Song, Chul-Hyun Cho, Ji Won Yeom, Taek Lee, Jung-Been Lee, Heon-Jeong Lee, and Jae Kyoung Kim, "Accurately Predicting Mood Episodes in Mood Disorder Patients Using Wearable Sleep and Circadian Rhythm Features, *npj Digital Medicine* 7 (2024), https://www.nature.com/articles/s41746-024-01333-z.

10. Tom Deboer, Mariska J. Vansteensel, László Détári, and Johanna H. Meijer, "Sleep States Alter Activity of Suprachiasmatic Nucleus Neurons," *Nature Neuroscience* 6 (2003): 1086–1090.

11. Daniel DeWoskin, Jihwan Myung, Mino D. C. Belle, et al., "Distinct Roles for GABA Across Multiple Timescales in Mammalian Circadian Timekeeping," *Proceedings of the National Academy of Sciences of the United States of America* 112 (2015): E3911–E3919.

Chapter 7

1. Alan L. Hodgkin and Andrew F. Huxley, "A Quantitative Description of Membrane Current and Its Application to Conduction and Excitation in Nerve," *Journal of Physiology* 117 (1952): 500–544.

2. David Paydarfar, Daniel B. Forger, and John R. Clay, "Noisy Inputs and the Induction of On-Off Switching Behavior in a Neuronal Pacemaker," *Journal of Neurophysiology* 96 (2006): 3338–3348.

3. Mino D. Belle, "Daily Electrical Silencing in the Mammalian Circadian Clock," *Science* 326 (2009): 281–284.

Chapter 8

1. Albert Goldbeter, *Biochemical Oscillations and Cellular Rhythms: The Molecular Bases of Periodic and Chaotic Behaviour* (Cambridge University Press, 1996).

2. John S. O'Neill and Akhilesh B. Reddy, "Circadian Clocks in Human Red Blood Cells," *Nature* 469 (2011): 498–503.

3. D. James Morre, Pin-Ju Chueh, Jake Pletcher, et al., "Biochemical Basis for the Biological Clock," *Biochemistry* 41 (2002): 11941–11945.

4. Satchin Panda, *The Circadian Code: Lose Weight, Supercharge Your Energy, and Tranform Your Health from Morning to Midnight* (Rodale, 2018).

5. Frank A. J. L. Scheer, Michael F. Hilton, Christos S. Matzoros, and Steven A. Shea, "Adverse Metabolic and Cardiovascular Consequences of Circadian Misalignment," *Proceedings of the National Academy of Sciences of the United States of America* 106 (2009): 4453–4458.

6. Jason E. McDermott, Jon M. Jacobs, Nathaniel J. Merrill, et al., "Molecular-Level Dysregulation of Insulin Pathways and Inflammatory Processes in Peripheral Blood Mononuclear Cells by Circadian Misalignment," *Journal of Proteome Research* 23 (2024): 1547–1558.

FURTHER READING

Bunning, Erwin. *The Physiological Clock*. Heidelberg Science Library, 1964.

Dunlap, Jay C., Jennifer J. Loros, and Patricia J. Decoursey. *Chronobiology: Biological Timekeeping*. Sinauer Associates, 2004.

Enright, Jim T. *The Timing of Sleep and Wakefulness*. Springer, 1980.

Forger, Daniel B. *Biological Clocks, Rhythms and Oscillations: The Theory of Biological Timekeeping*. MIT Press, 2017.

Foster, Russell. *Life Time: Your Body Clock and Its Essential Roles in Good Health and Sleep*. Yale University Press, 2022.

Glass, Leon, and Michael C. Mackey. *From Clocks to Chaos: The Rhythms of Life*. Princeton University Press, 1988.

Goldbeter, Albert. *Biological Oscillations and Cellular Rhythms*. Cambridge University Press, 1996.

Panda, Satchin. *The Circadian Code*. Rodale Books, 2018.

Strogatz, Steven H. *The Mathematical Structure of the Human Sleep-Wake Cycle*. Springer, 1986.

Winfree, Arthur. *The Timing of Biological Clocks*. Scientific American Library, 1987.

INDEX

DANIEL FORGER is the Browne Professor of Science, Professor of Mathematics, and Research Professor of Computational Medicine and Bioinformatics at the University of Michigan, where he directs the Michigan Center for Applied and Interdisciplinary Mathematics. Dr. Forger is also the CSO and a founder of Arcascope, a company that makes circadian rhythms software.

Publisher contact:
The MIT Press
Massachusetts Institute of Technology
77 Massachusetts Avenue, Cambridge, MA 02139
mitpress.mit.edu

EU Authorised Representative:
Easy Access System Europe, Mustamäe tee 50,
10621 Tallinn, Estonia
gpsr.requests@easproject.com

Printed by Integrated Books International,
United States of America